Ball Culture Guide

The Encyclopedia of Seed Germination (2nd Edition)

by Jim Nau

Ball Publishing

Batavia, Illinois USA

Ball Publishing
335 North River Street
P.O. Box 9
Batavia, IL 60510-0009 USA

Library of Congress Cataloging in Publication Data

Nau, Jim, 1958-
    Ball culture guide : the encyclopedia of seed germination / Jim
Nau. -- 2nd ed.
      p.   cm.
    Includes bibliographical references and index.
    ISBN 1-883052-01-7 : $39.00
    1. Plants, Ornamental--Propagation--Handbooks, manuals, etc.
2. Vegetables--Propagation--Handbooks, manuals , etc.   3. Plants,
Ornamental--Seeds--Handbooks, manuals, etc.   4. Vegetables--Seeds-
-Handbooks, manuals, etc.   5. Sowing--Handbooks, manuals, etc.
6. Germination--Handbooks, manuals, etc.   I. Title.
SB406.83.N83   1993
635' .0431--dc20   93-31746
                                CIP

Cover photo: 1991 Field Trials, Ball Seed Co.

CONTENTS

In Memory
My parents were always there to encourage me to get the most that I could out of life. I have my mother's attention to detail and my father's perseverance to get the job done. But most of all I have their humor. After all, you have to have a sense of humor when your parents are named Dick and Jane. My life is rich and full but a little less so without their presence.

Acknowledgments
Special thanks to Anne Brennan, book production intern at Ball Publishing, for countless hours of editing and proofreading to get this second edition on its way. Thanks also to Paul Pilon, vegetable trials intern at Ball Seed Co., for researching plant nomenclature changes.

The Starting Point

For easy reference, the *Ball Culture Guide* is divided into six sections. Some crops, such as African marigolds and zinnias, are in two separate sections based on how they are to be grown. An index at the back of the *Guide* will help you locate a crop quickly. Please keep in mind, however, that culture will vary from one grower to the next, depending on temperatures, media, light and other factors.

You will need to try the crop yourself and maintain your own records to determine if a particular crop is suitable for you. Just remember that if you change any facet of the culture, it will have an effect on the finished crop.

Information in the *Ball Culture Guide* is intended to provide you with a starting point for growing new or unfamiliar crops, and I hope these basic guidelines prove useful to you.

Crop Types

The Bedding, Florist and Foliage Plants section differs from the other five sections, in that it is a composite of many types of crops–primarily annuals. However, a number of other types of crops are also listed here, even cut flowers and perennials, since many have dual markets. If a plant is used more often as an annual because of its foliage or flower qualities rather than its perennial or culinary traits, we have placed it in the Bedding, Florist and Foliage Plants section with the following codes:

A   Annuals
P   Pot Plants
C   Cut Flowers
F   Foliage Plants
Pe Perennials

Germination Lighting

L indicates seed should be exposed to light while germinating.

C means seed should be covered with media as the directions indicate.

C. Lt. requires seed to be lightly covered, leaving the seed as close to the surface of the media as possible. Some of these varieties perform better if left exposed to light, but need a thin covering to maintain humidity and moisture. Germination tips and techniques vary from publication to publication, depending on the latest research developments from universities and private firms.

The standards in the horticulture trade are based on work originally done by Dr. Henry (Marc) Cathey of the National Arboretum. A wide range of plant material was included in his research, however, not all classes could be tested. Therefore, culture information varies from source to source for those classes that were not assigned standards.

Growing Temperatures and Crop Times

These refer to **night temperatures** once plants are established in final packs or containers. Plants must be strong, actively growing seedlings before the warm temperatures required for germination are dropped down to the cooler temperatures (68° F/19° C) needed for growing on. Problems can develop later from dropping temperatures too early on seedlings that have not yet established roots in new containers after being transplanted from germination plug trays or seed flats. Use your own judgment on the right time to drop temperatures for finishing each crop.

The number of weeks indicated for finishing crops green or in flower are based on growing in standard 11 x 22-in. flats of 32 cells unless otherwise stated. When using smaller cell packs, such as 48s or 72s, crop times can differ by as much as 1 to 2 weeks. Keep in mind that the smaller the volume of root space, the faster the crop will develop and become salable. Unless otherwise stated, these times refer to the normal growing season for each crop.

Bedding plants are sown and finished in the spring, while cool pot crops are grown during the fall and early winter months. Our own trial notes, taken over the past 10 years, provide the basis for most crop times noted in this guide. Growing seasons that are warmer or colder than average can change crop times dramatically.

Growth Regulators

This information is provided only for the Bedding, Florist and Foliage Plants section. At the time of this printing, few growth regulators are labeled for use on perennial crops.

Growth regulators are classified as pesticides. The container label will explain how the chemical should be applied and any possible consequences, such as late flowering, foliage damage or other problems that can occur with their use. Always read and follow label instructions. The information provided in the *Ball Culture Guide* is intended to be used only as a guideline. We do not take responsibility for the use of any growth regulator on any crop.

In the chart for each crop you will find numbers indicating which growth regulators may be used. The numbers below refer to chemicals labeled by the manufacturer for use on the crop:

1 A-Rest
2 B-Nine
3 Bonzi
4 Cycocel

The following numbers reflect the chemicals that have been used by growers and university researchers to keep crops dwarf, *but have not been cleared by the manufacturer for use on the crop.* They are noted only for your information:

5 A-Rest
6 B-Nine
7 Bonzi
8 Cycocel

Interpreting Garden Location

F. Sun means the variety should be planted to a garden in full sunlight.

P. Shade indicates that a bed with partial shade and partial sun is needed. This usually applies to classes that perform well in shade but require bright light to grow and flower.

There is a wide range of shade-loving plants that perform well even when grown in sunny locations. The trick is to plant them closer together in the garden–6 to 8 in./15 to 20 cm. apart–with sun from only morning until noon. If they're in the shade during the hottest part of the day, these plants will put on a good show. Remember, however, that shade-loving plants placed in bright areas will initially require more water after planting.

Tender/Hardy Plants

H indicates plants that can survive cool weather, including pansies and snaps as well as other cool-weather annuals and tender perennials.

T refers to tender crops that require warm temperatures to grow and, therefore, do not perform well when temperatures drop below 50° F (10° C) for a long period of time.

Culture Comments

For some crops you will find basic culture techniques specific to the class, along with variety recommendations and suggested uses. Where appropriate, varieties suitable for mass plantings in landscape settings are listed.

Because the climate and conditions for southern growers vary greatly from the Chicago latitude, we have included specific culture for this region throughout the Bedding, Florist and Foliage Plants section when information was available. "South" refers to the southernmost tier of states from Texas to North Carolina, specifically USDA Zones 8 and 9. Due to the subtropical conditions, Zone 10 (Miami and southern Texas) is the only area of the United States for which no culture is provided.

Cut Flowers Overview

The charts contain the basic requirements for producing each crop, whether as cut flowers or as bedding plants. To grow specifically for cut flower production, see culture comments for greenhouse growing and field growing.

Herbs and Vegetables Overview

The count in the Days to Maturity column reflects the time from planting to the garden until first picking. For a total crop time from sowing to first picking, use the cell pack crop time plus the Days to Maturity.

Overview on Perennial Chapter

Germination needs and growing techniques vary greatly among perennials. The methods available to force plants into bloom the same season from seed are indicated for each crop; however, use of these methods does not guarantee plants will be in flower when sold. Instead, blooming will be induced between spring and fall of the same season plants are sold. The methods for forcing are defined as follows:

Method 1: All perennials listed under this classification and its categories are sown from spring through fall to build up strong plants for over-wintering. These plants should flower the following year. Method 1 includes many spring-flowering perennials such as aquilegia and Oriental poppy. However, perennials that bloom within several months of sowing can also be started the previous year in order to produce larger plants for spring sales.

All perennials listed in this category are best treated in one or more of the following ways:

A) Sow from June to August for transplanting to quart (.95 l.) or gallon (3.8 l.) containers before the middle of September in the Chicago area. It is important to have the root systems developed throughout the soil volume prior to over-wintering. Once the roots are established, plants should be over-wintered until spring in an unheated greenhouse or cold frame with a soil temperature around 30° to 35° F (-1° to 1° C). Slower germinating classes, such as armeria, may require an earlier sowing–about April– so that plants will be strong by fall.

B) Biennial crops produce a rosette of foliage the first year and flower the second year after they have undergone a cold period to initiate flower buds. Sow seed from June to August, then continue as in Method 1A. Alternatively, sow in

the spring or early summer for fall sales of green 4- and 6-in. (10- and 15-cm.) pots along with bulbs. Sales should end approximately 5 to 6 weeks before the first expected frost, allowing plants to become established in soil before winter. If a biennial is sown between January and March and transplanted as necessary until sold, flowering will generally be delayed until the following year. However, if the plants are grown under stress, such as rootbound conditions or insufficient chilling, they may bolt and flower the same season. If this happens, many varieties of biennials will not flower again, and the plant should be thrown away or dug from the garden. Be sure to instruct home gardeners to place a minimum of 2 in. (5 cm.) of mulch at the base of all fall-planted perennials after several frosts to protect them against the heaving that could kill plants in the spring.

C) In a variation of Method 1A, sow seed in late summer and early fall for strong plants ready for transplanting from large, 18-cell packs or quart (.95 l.) containers by Christmas. (Digitalis and several other crops grow quickly and require 6-in. (15-cm.) pots by the end of December.) Varieties in this classification are grown cool at 40° F (4° C) until they are sold in the spring. Under this treatment, some spring-flowering plants will be in bloom by March. Method 1C particularly benefits slow-germinating and sensitive seed which often cannot take the heat of summer. It also works well for seasonal nurseries which close down during the summer months. However, keep in mind that perennials grown during the fall will compete for greenhouse space with your fall and winter holiday crops.

Method 2: Plants in this classification are generally warm-season perennials which flower during the summer and early fall.

Perennials Overview

Sow from January to March and transplant to 4- and 6-in. (10- and 15-cm.) pots (quart/.95 l. and gallon/3.8 l. containers). Plants will sometimes flower in the garden during the summer. As plants become established in the garden, subsequent seasons will produce larger and more free-flowering plants.

For blooming sales from April to June, sow seed from late October to early December and grow on at the temperatures noted in the culture chart. Some trial and error will be required, however, to determine the exact conditions necessary to promote flowering of each crop. Within the same class, for example, two species grown at the temperature noted for that crop may reach the flowering stage at different times. Be sure to keep accurate notes.

Note: Plants forced into out-of-season bloom for flowering sales may be out of bloom during their natural flowering period.

Method 3: One of the older methods for growing perennials is to sow them from June to August and transplant into cold frames or the field for spring digging. Plants are sold in gallon (3.8 l.) containers in the spring.

Over-Wintering Comments: Regardless of the size, never over-winter perennials in packs since individual plants will not emerge uniformly from dormancy. Watering and feeding begun on flats with plants in various stages of development can lead to losses from diseases, over-watering and/or over-feeding.

Ornamental Grasses Overview

A new addition to the *Culture Guide* is the section on ornamental grasses. Though a number of selections are available, this guide doesn't cover the multitude of individual varieties and cultivars. It does, however, provide some basic information on the propagation of some of the more common types available in the trade from seed.

A key point is that all of these grasses resent individual seedling transplanting. Yes, it can be done, but it often slows the development of the plant and can add up to 3 weeks to the crop time. Keep in mind that the crop times below are mostly for green plants, since a full flowering plant can be very large at the time of bud set.

Finally, remember that annual varieties will always fill a flat more quickly than perennial ones. From the time of seeding until close to green pack sales, the perennial and tender perennial grasses often appear less vigorous than their annual seeded counterparts.

Crop Types

A	Annual
T. Pe	Tender Perennial
Pe	Perennial
C	Cut Flower

Reference Sources

A-Rest (label). 1988. Indianapolis, IN: Elanco Products.

Armitage, Allan M. 1989. *Herbaceous Perennial Plants.* Athens, GA: Varsity Press.

Ball RedBook. 14th ed. 1985. Ed.Vic Ball. Reston, VA: Reston Pub. Co., Inc.

Ball RedBook. 7th, 8th, 10th, and 11th ed. 1948-1965. Ed. Vic Ball. Aurora, IL: Kelmscott Press, Inc.

Ball Seed Catalog. 1968-1988. West Chicago, IL: Ball Seed Co.

Ball Seed Co. Trial Notes. 1984-1988. West Chicago, IL: Ball Seed Co.

Benary Seed Co. *Hardy Perennials from Seed.* Hann Muenden, Germany: Benary Seed Co.

Bonzi (label). 1988. Des Plaines, IL: Sandoz Crop Protection Corp.

Cycocel (label). 1988. Wayne, NJ: American Cyanamid Corporation.

Foliage Germination and Growing. (Super Seedlings cultural instructions). Parrish, FL: Pan American Seed Co.

Hortus Third. 1976. New York: Macmillan Pub. Co., Inc.

McCollum, J.P. 1980. *Producing Vegetable Crops.* 3rd ed. Danville, IL: Interstate Printers and Pub.

Perry, L.P., and J.W. Boodley. *Growing Foliage Plants from Seed.* Ithaca, NY: Cornell University, Dept. of Floriculture and Ornamental Horticulture.

Pinnell, M.M., A.M. Armitage, and D. Seaborn. 1985. *Germination Needs of Common Perennial Seed.* Research Bulletin No. 331. The University of Georgia, College of Agriculture Experiment Station.

Post, K. 1952. *Florist Crop Production and Marketing.* New York: Orange Judd Pub. Co.

Royal Sluis Seed Co. *Growing Cut Flowers from Seed.* Enkhuizen, Holland: Royal Sluis Seed Co.

Still, Steven. 1988. *Herbaceous Ornamental Plants.* Champaign, IL: Stipes Pub. Co.

USDA Plant Hardiness Zone Map

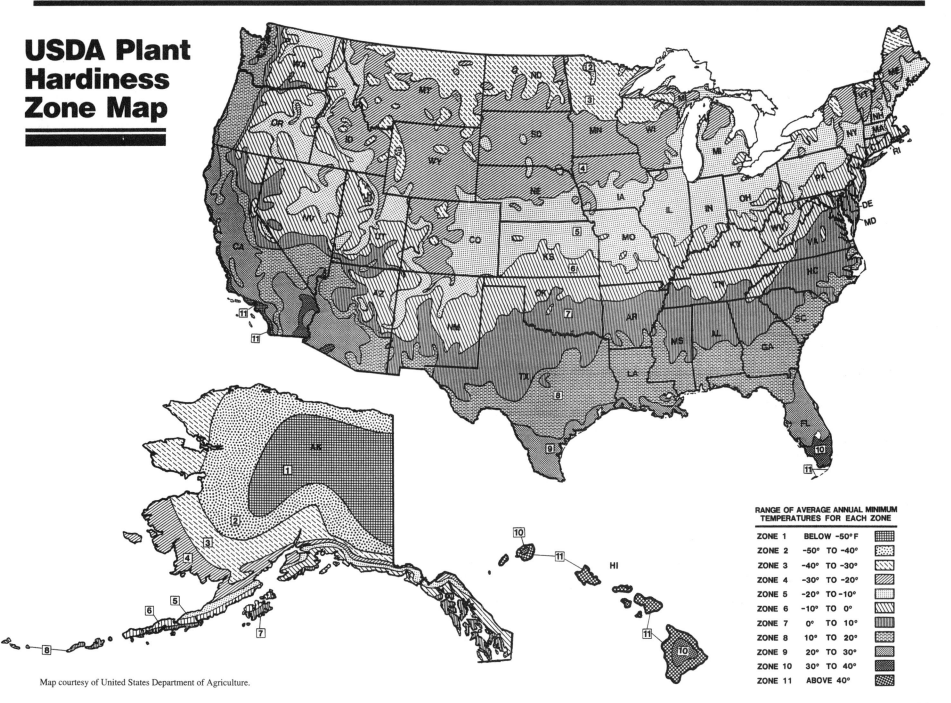

Map courtesy of United States Department of Agriculture.

RANGE OF AVERAGE ANNUAL MINIMUM TEMPERATURES FOR EACH ZONE

ZONE 1	BELOW -50°F	
ZONE 2	-50° TO -40°	
ZONE 3	-40° TO -30°	
ZONE 4	-30° TO -20°	
ZONE 5	-20° TO -10°	
ZONE 6	-10° TO 0°	
ZONE 7	0° TO 10°	
ZONE 8	10° TO 20°	
ZONE 9	20° TO 30°	
ZONE 10	30° TO 40°	
ZONE 11	ABOVE 40°	

BEDDING, FLORIST & FOLIAGE PLANTS

Class	Type	Seed for 1,000 plants (oz.)	Number of Seeds	Germination Temperature Fahrenheit	Germination Temperature Celsius	Lighting	Days To Germinate	Days Sowing To Transplant	Growing On Temperature Fahrenheit	Growing On Temperature Celsius
ABELMOSCHUS A. moschatus (Annual Hibiscus)	A	1/2	2,850/oz. 100/g.	75°	24°	—	10–14	15–22	60°–65°	15°–18°

Ball Culture: Treat abelmoschus as you would vinca or hibiscus. To sow, leave seed uncovered and place flats in darkness, or cover seed lightly and keep away from bright, direct light. Grow on with warm night temperatures of no less than 60° F (15° C).

Like hibiscus, the flowers of abelmoschus last only 1 day. Though the plants will fill out the containers, the short life-span of the blooms means abelmoschus will not be as free–flowering as you prefer for hanging baskets.

As a landscape item, abelmoschus is best used in containers rather than mass plantings, except in warm southern locations. We recommend trying this variety in small quantities the first year.

Class	Type	Seed for 1,000 plants (oz.)	Number of Seeds	Germination Temperature Fahrenheit	Germination Temperature Celsius	Lighting	Days To Germinate	Days Sowing To Transplant	Growing On Temperature Fahrenheit	Growing On Temperature Celsius
AFRICAN VIOLET Saintpaulia ionantha	P	1/128 = 5,000 seeds	1,000,000/oz. 35,000/g.	70°	21°	L	18–25	65–80	65°	18°

Ball Culture: Crop times listed above are based on the newer strains which show marked improvement over the older varieties, including earlier flowering and a wider selection of colors. Older varieties, such as Concerto, require anywhere from 40 to 52 weeks to flower, with half of the crop time spent in the seedling tray.

Class	Type	Seed for 1,000 plants (oz.)	Number of Seeds	Germination Temperature Fahrenheit	Germination Temperature Celsius	Lighting	Days To Germinate	Days Sowing To Transplant	Growing On Temperature Fahrenheit	Growing On Temperature Celsius
AGERATUM A. houstonianum	A, C	1/128	200,000/oz. 7,000/g.	78°–82°	25°–28°	L	8–10	15–20	60°–65°	15°–18°

Ball Culture: For the best pot performance, hold in 2¼-in. (5.5-cm.) pots (cell packs) until buds form, then transplant. In the garden, plants can be spaced 12 in. (30.5 cm.) apart, or 10 in. (25.5 cm.) apart for a fuller effect. Heights for dwarf varieties generally range from 6 to 9 in. (15 to 23 cm.), with exceptions such as Adriatic and Blue Blazer which reach up to 14 in. (35.5 cm.) in the garden.

In the southern U.S.: Allow 9 to 10 weeks for flowering pack sales and 11 to 12 weeks for flowering 4-in. (10-cm.) pot sales of dwarf varieties. Sales should begin once all danger of frost has passed and continue through April for flowering through mid-August. Plant in full sun to partial shade, and space 6 to 10 in. (15 to 25.5 cm.) apart.

Recommended varieties: Blue Puffs (also called Blue Danube) and Blue Hawaii are the standards for mid-blue flowers, the most popular color for landscaping, followed by purple. White ageratums are also available in limited supply.

Class	Type	Seed for 1,000 plants (oz.)	Number of Seeds	Germination Temperature Fahrenheit	Germination Temperature Celsius	Lighting	Days To Germinate	Days Sowing To Transplant	Growing On Temperature Fahrenheit	Growing On Temperature Celsius
ALYSSUM, SWEET Lobularia maritima	A	1/16*	90,000/oz. 3,150/g.	78°–82°	25°–28°	L	8–10	**	50°–55°	10°–13°

Ball Culture: In the northern U.S., alyssums may suffer heat stall during the hottest part of the summer and some may die out. However, white varieties are more persistent and will reflower once the night temperatures cool.

In the southern U.S.: Allow 7 weeks for early flower color in packs and 9 weeks for 4-in. (10-cm.) pots. Plant late summer to mid winter for plants that will flower up to June. The white varieties may flower year-round. In the garden, plants do best in full sun to partial shade, spaced 8 in. (20 cm.) apart. Alyssums are hardy only in mild-winter areas and will freeze out at 27° F (-3° C) and below.

Recommended varieties: New Carpet of Snow is the standard white variety, sweetly scented and early to flower. Snow Crystal, however, offers the largest flower of the white varieties. For uniform habit and excellent outdoor performance, the Wonderland series is ideal. The 3 separate colors are well-matched in crop time and habit, with the aptly-named Wonderland Deep Rose among the best

	Crop Time (Weeks)			No. Plants							
Green Packs	Flower Packs	Pots 4-in./10-cm.	Baskets 10-in./25.5-cm.	Pot 4-in./10-cm.	Basket 10-in./25.5-cm.	Pinching	Growth Regulators	Garden Height	Staking	Location	Tender/ Hardy
9–11	12–13	14–16	17	1	4–5	*	8	12–15 in. 30.5–38 cm.	No	F. Sun	T

*As it's growing, this crop requires plenty of light to branch. Under the short days of spring, plants tend to grow upright without branching; pinching out the tip will encourage branching but will also increase the crop time.

—	—	25–27	—	1	—	N	—	—	—	—	T

8–9	10–11	12–13	—	1	—	No	1, 2, 3	6–9 in. 15–23 cm.	No	F. Sun	T

There are several cut flower varieties similar in color to Blue Puffs. They flower several weeks later than the dwarf varieties and grow 2 to 2½ ft. (61 to 76 cm.) tall.

7	8–9	10–11	12–14	—	—	No	—	3–6 in. 7.5–15 cm	No	F. Sun to P. Shade	H

rose flower colors on the market. All of these varieties work well in landscapes on 10- to 12-in. (25.5- to 30.5-cm.) centers.

*Based on multiple-sown seed.

**Sow direct to the final container using a number of seeds to provide several seedlings in each cell or 4-in. (10-cm.) pot. White flowered varieties are the most vigorous of the colors, and the most heat tolerant.

Class	Type	Seed for 1,000 plants (oz.)	Number of Seeds	Germination Temperature Fahrenheit	Celsius	Lighting	Days To Germinate	Days Sowing To Transplant	Growing On Temperature Fahrenheit	Celsius
AMARANTHUS A. tricolor	A	1/32	44,000/oz. 1,540/g.	70°–75°	21°–24°	C. Lt.	8–10	18–20	65°–68°	18°–20°

Ball Culture: Plants grow quickly and are sold for their brightly-colored foliage instead of their flowers. Amaranthus can be used as bedding plants or cut flowers. Growing upright with little branching, plants tend to get top-heavy, so some staking is needed. After 7 weeks, standard varieties such as Flaming Fountains or Early Splendor can easily reach 9 to 12 in. (23 to 30.5 cm.) in the pack. Amaranthus will not tolerate wet soils, preferring warm, dry locations, so plant only in well-drained areas, spaced 12 to 15 in. (30.5 to 38 cm.) apart. Amaranthus is tender and will not tolerate a frost.

Class	Type	Seed for 1,000 plants (oz.)	Number of Seeds	Germination Temperature Fahrenheit	Celsius	Lighting	Days To Germinate	Days Sowing To Transplant	Growing On Temperature Fahrenheit	Celsius
ANCHUSA A. capensis	A,P	1/4	10,000/oz. 380/g.	68°–72°	20°–21°	L	4–8	15–25	55°–60°	13°–15°

Ball Culture: Crop is actually a biennial best treated here as an annual. Early April sowings are salable green by early June and will start to flower by the first week of July. Once flowers begin to open, plants my take up to two weeks to achieve full bloom. Anchusa produces medium- to azure-blue flowers on plants that grow relatively quickly though are somewhat variable in height and habit. In the pack, as the roots reach the sides of the container, the lowest leaves often turn yellow. Stress caused by under-fertilizing and cool, cloudy conditions appear to accentuate this condition.

Class	Type	Seed for 1,000 plants (oz.)	Number of Seeds	Germination Temperature Fahrenheit	Celsius	Lighting	Days To Germinate	Days Sowing To Transplant	Growing On Temperature Fahrenheit	Celsius
ARALIA Dizygotheca elegantissima	F	1/4	9,500/oz. 332/g.	Day: 85° Night: 68°	Day: 29° Night: 20°	L	35–42	70–85	70°–85°	21°–29°
A. Sieboldii (Fatsia japonica)	F	1/4	5,000/oz. 175/g.	Day: 85° Night: 68°	Day: 29° Night: 20°	L	28–40	56–70	60°–75°	15°–24°

Ball Culture: Reduce germination temperatures for both Dizgotheca and Sieboldii varieties at night. For foliage sales of Dizygotheca, allow 27 to 30 weeks for 4-in. (10-cm.) pots, with 1 plant per pot. For 6-in. (15-cm.) pots with 3 to 4 plants each, allow 32 to 35 weeks. This type of aralia is best suited as an upright-growing specimen plant.

Class	Type	Seed for 1,000 plants (oz.)	Number of Seeds	Germination Temperature Fahrenheit	Celsius	Lighting	Days To Germinate	Days Sowing To Transplant	Growing On Temperature Fahrenheit	Celsius
ASPARAGUS A. densiflorus cv. Myers	F	3**	500/oz. 17/g.	Day: 85° Night: 75°	Day: 29° Night: 24°	C	21–42	—	65°–70°	18°–21°
A. setaceus	F	3**	875/oz. 30/g.	Day: 85° Night: 75°	Day: 29° Night: 24°	C	21–42	—	65°–70°	18°–21°
A. densiflorus cv. Sprengeri	F	3**	650/oz. 23/g.	Day: 85° Night: 75°	Day: 29° Night: 24°	C	21–42	—	65°–70°	18°–21°

Ball Culture: For cv. Myers, allow 28 to 32 weeks for 4-in. (10-cm.) pots with 1 plant each, and 36 to 40 weeks for 10-in. (25.5-cm.) baskets with several plants. For plumosus varieties, allow 22 to 26 weeks for 4-in. (10-cm.) pots with 1 plant per pot, and 26 to 30 weeks for 6-in. (15-cm.) pots with up to 3 plants per pot. Sprengeri varieties require 22 to 26 weeks for 4-in. (10-cm.) pots with 1 plant each, and 30 to 34 weeks for 10-in. (25.5 cm.) baskets with several plants. **Most often sold by seed count rather than by weight.

Crop Time (Weeks)				No. Plants		Pinching	Growth Regulators	Garden Height	Staking	Location	Tender/ Hardy
Green Packs	Flower Packs	Pots 4-in./10-cm.	Baskets 10-in./25.5-cm.	Pot 4-in./10-cm.	Basket 10-in./25.5-cm.						
6–7	—	8–9	—	1	—	No	—	2.5–3 ft. 76–91.5 cm.	Yes	F. Sun	T

In the southern U.S.: Once all danger of frost has passed, plants can be sold until May for colorful foliage throughout the summer. Plants can reach 3½ to 4 ft. (1.1 to 1.2 m.) in height.

Since our experience has shown amaranthus isn't reliable in displays and requires extra time to look its best, we recommend trying this crop in limited amounts for landscaping the first season.

Crop Time (Weeks)				No. Plants		Pinching	Growth Regulators	Garden Height	Staking	Location	Tender/ Hardy
7–10	—	—	—	1	—	No	—	7–12 in. 17.5–25.5 cm.	No	F. Sun to P. Shade	T
—	—	—	—	—	—	No	—	—	—	—	T
—	—	—	—	—	—	No	—	—	—	—	T

Sieboldii will grow faster than Dizygotheca; allow 17 to 20 weeks for a 4-in. (10-cm.) pot with 1 plant, or 23 to 25 weeks for a 10-in. (25.5-cm.) basket with 4 to 6 plants.

This type is ideal for baskets, and performs equally well in 6-in. (15-cm.) pots.

Crop Time (Weeks)				No. Plants		Pinching	Growth Regulators	Garden Height	Staking	Location	Tender/ Hardy
—	—	—	—	1	3	No	—	—	No	F. Sun to P. Shade	T
—	—	—	—	1	3	No	—	—	No	F. Sun to P. Shade	T
—	—	—	—	1	3	No	—	—	No	F. Sun to P. Shade	T

Class	Type	Seed for 1,000 plants (oz.)	Number of Seeds	Germination Temperature Fahrenheit	Celsius	Lighting	Days To Germinate	Days Sowing To Transplant	Growing On Temperature Fahrenheit	Celsius
ASTER **Callistephus chinensis** *Also see Cut Flowers and Perennial Plants sections.*	A, P	1/8	12,000/oz. 420/g.	70°	21°	—	8–10	15–20	60°–62°	15°–17°
BALSAM **Impatiens balsamina**	A	1/2	3,300/oz. 115/g.	70°	21°	C. Lt.	8–10	20–25	60°–62°	15°–17°
BEGONIA, FIBROUS **B. x semperflorens-cultorum** **(Wax begonia)**	A	*	2,000,000/oz. 70,000/g.	78°–80°	25°–26°	L	14–21	45–50	60°	15°

ASTER
Callistephus chinensis

Ball Culture: For spring flowering sales, asters require long days. Plants need an additional 4 hours of light each day, from 1 week after sowing until color shows or May 5—whichever comes first—based on a 60° F (15° C) night temperature.

For pot sales, allow 4 to 4½ months to finish from February to May sowings, and 5 to 5½ months for November and December sowings. Plants will flower in the summer.

In the southern U.S.: Have plants ready for sale after all danger of frost has passed, until April. For bedding plant sales, sell green in packs. Dwarf varieties should be planted in full sun to light shade, spaced 10 to 12 in. (25.5 to 30.5 cm.) apart.

BALSAM
Impatiens balsamina

Ball Culture: Balsams prefer to be treated like their relatives, the impatiens. In the garden, however, balsams grow more upright and tolerate more sun than impatiens. Since the flowers hide within the foliage, it's difficult to use this

annual for anything more than a bedding plant. Balsams become tall and spindly if allowed to flower in the pack, so sell green or in flower in 4-in. (10-cm.) pots.

In the southern U.S.: Plants do best in full sun to partial shade, and can reach up to 2 ft. (61 cm.) in height in the warmest areas. After all danger of frost has passed, plant to the garden until April. Balsams should be spaced 12 to 15 in. (30.5 to 38 cm.) apart.

BASIL, ORNAMENTAL
Ocimum basilicum
(Dark Leaf Basil)
See Herbs section.

BEGONIA, FIBROUS
B. x semperflorens-cultorum
(Wax begonia)

Ball Culture: Recent research has shown that the highest germination rates for this crop are obtained by sprinkling seed evenly over medium and germinating at 78° to 80° F. If this is impractical, use a constant temperature of 70° to 72° F (21° to 22° C) throughout germination.

*Most often sold by seed count rather than by weight.

This is one of the few crops in which plants bloom before they are large enough to be sold and handled in the garden. Begonias are not frost tolerant, and should be planted after all danger of frost has passed. Dwarf series such as Prelude, Varsity and Scarletta require 10 to 12 in. (25.5 to 30.5 cm.) garden spacing. Taller series such as Avalanche,

Encore, Party and other large-leaved, large-flowered varieties should be spaced at 12 to 14 in. (30.5 to 35.5 cm.).

Both green and bronze-leaved varieties are available. Green-leaved varieties will darken somewhat if planted in full sun, but the color will not approach that of their bronze-leaved counterparts. In full sun locations, plants need to be established before the weather turns hot, otherwise they are slow to develop and fill in. When planted late, flowers and foliage tips burn under hot, dry conditions, so fibrous begonias should be planted in afternoon shade at this point.

In the southern U.S.: Allow 14 to 15 weeks for flowering packs, and 16 to 17 weeks for flowering 4-in. (10-cm.) pots. In the garden, space dwarf series 8 to 10 in. (20 to 25.5 cm.) apart in full sun or lightly shaded areas. Begonias will flower longer if planted in shade rather than direct sun.

Recommended varieties: In tall, upright types, the Party and Encore series are the best overall varieties. Plants grow to 12 in. tall with 1½-in. flowers. Encore includes more flower colors than Party. Dwarf, compact types such as the Prelude, Varsity, Olympia and Cocktail series predominate, offering excellent overall performance in gardens, landscape plantings and containers. The Preludes do best as border, bedding or 4-in. (10-cm.) pot plants, and along with

Crop Time (Weeks)				No. Plants							
Green Packs	Flower Packs	Pots 4-in./10-cm.	Baskets 10-in./25.5-cm.	Pot 4-in./10-cm.	Basket 10-in./25.5-cm.	Pinching	Growth Regulators	Garden Height	Staking	Location	Tender/ Hardy
7–8	—	15–16	—	1	—	No	1, 2	8–12 in. 20–30.5 cm.	No	F. Sun	T

Recommended varieties: Pixie Princess Mixture is our best-selling bedding and 4-in. (10-cm.) pot aster, standing 8 in. (20 cm.) tall with fully double blooms up to 3 in. (7.5 cm.) across. Another favorite, Dwarf Queen Mixture, grows up to 12 in. (30.5 cm.) tall with 2½-in. (6-cm.), fully double blooms. Our popular Pot 'n Patio series stands 6 in. (15 cm.) tall and is available in several separate colors plus a mixture. It flowers without additional lighting, so sowings

made in mid-January will be in flower by mid-April. Seedlings should be transplanted directly from seedling trays to the final containers. Pot 'n Patio flowers under stress, especially if it becomes root-bound in the container as in cell packs. We recommend selling this variety in pots only.

Asters are not a popular choice for landscaping due to their short flower season and susceptibility to Asters Yellow disease transmitted by leaf hoppers. See the Cut Flowers section for detailed culture information on cut flower varieties.

7–8	—	—	—	—	—	No	1	12–16 in. 30.5–41 cm.	No	F. Sun to P. Shade	T

14–15	16	18	19–20	1–2	5–6	No**	6	6–10 in. 15–25.5 cm.	No	F. Sun to P. Shade	T

the Varsities and Olympias are our earliest flowering strains. For best results, space them 8 to 10 in. (20 to 25.5 cm.) apart. Varsities and Olympias are more vigorous than Preludes, and have the largest flower of any of our dwarf series. Spaced 10 in. (25.5 cm.) apart, they create a striking effect in landscapes and beds. Among the dark-leaved varieties, the Cocktail and Bingo! series are ideal choices for landscaping, bedding or container plantings, though the Bingos! are more vigorous.

**Pinching isn't necessary for the dwarf varieties, but we recommend a soft pinch (removing the terminal) on taller fibrous types such as Encore and Party. These series tend to grow upright, with minimal basal branching.

BEDDING, FLORIST & FOLIAGE PLANTS

BEDDING, FLORIST & FOLIAGE PLANTS

Class	Type	Seed for 1,000 plants (oz.)	Number of Seeds	Germination Temperature		Lighting	Days To Germinate	Days Sowing To Transplant	Growing On Temperature	
				Fahrenheit	Celsius				Fahrenheit	Celsius
BEGONIA, TUBEROUS B. x tuberhybrida	A	*	1,000,000/oz. 35,000/g.	75°–78°**	24°–25°	L	15–30	45–50	60°	15°
BRACHYCOME B. iberidifolia (Swan River Daisy)	A	1/64	158,000/oz. 5,600/g.	70°	21°	L	4–8	14–21	55°–60°	13°–15°
BROWALLIA B. speciosa	A	1/64	125,000/oz. 4,375/g.	75°	24°	L	7-15	40–45	60°–65°	15°–18°
CALCEOLARIA C. herbeohybrida	P	—	500,000– 1,000,000/oz. 17,500–35,000/g.	70°	21°	L	10–16	24–32	50°–55°	10°–13°

Ball Culture: Sprinkle seed evenly over growing medium. Tuberous begonias from seed, such as Nonstops, Clips and Musicals, require long, 12-hour days from October to March. Add lights from 10 p.m. until 2 a.m., or light crops for 4 hours after sunset.

The number of blooms desired determines the length of the crop time. We have had blooming 10-in. (25.5-cm.) baskets of Nonstops in just 17 weeks. Typically in our trials, several flowers per plant have appeared on 10-in. (25.5-cm.) baskets 17 to 18 weeks after sowing, with loads of blooms after 23 to 24 weeks.

Though they burn in full sun, tuberous begonias perform well in morning sun with afternoon shade. Space on 10 to 12 in. (25.5 to 30.5 cm.) centers in the garden. Remember that when a tuberous variety begins to bloom, the double male flowers appear first, followed by the single female flowers.

In the southern U.S.: Allow 18 to 19 weeks for flowering 3- to 4-in. (7.5- to 10-cm.) pots, 20 to 21 weeks for flowering 10-in. (25.5-cm.) baskets with 4 to 5 plants each. Begonias are not frost tolerant. After all danger of frost has passed,

Comments: Although sometimes sold as a mixture, separate colors of Brachycome are also available. These include blue or white, though off-types or shades will be seen in each color. Separate colors of yellow, pink, and lavender are also available from cuttings. Flowers are single to semi-double and measure less than 1 in. across. Brachycome is basal-branching and has a bushy habit in the garden. The stem itself is wiry which allows the plant to shift easily in the wind; this often creates an uneven performance in flower beds when used in mass plantings.

Brachycome works best in sunny-morning locations as a

Ball Culture: Carefully harden plants off before selling for outdoor use. Browallias tend to perform better in landscapes when planted out in full bloom from packs or pots. This crop is also ideal for hanging baskets on patios or suspended from the lowest branches of tall trees. In 10-in. (25.5-cm.) baskets, browallias can easily get up to 2 ft. (61 cm.) wide and 15 in. (41 cm.) tall. Though plants burn in full sun, they do well in morning sun with afternoon shade.

In the southern U.S.: Browallias will overwinter in areas without frost. Plant from mid-September to November in the deep South for flowering until late April. In cooler areas, have plants ready for sale once the danger of frost has passed.

Ball Culture: Sow from early August to mid-September for blooming plants from early January through mid-April. Keep soil temperatures above 70° F (21°C) during germination, and avoid overwatering once seedlings are up. Transplant to 2 1/4-in. (5.5-cm.) pots (cell packs) and grow on at 55° to 58° F (13° to 14° C) night temperatures once they are established in cell packs. Shift to final containers and establish at 60° F (15° C) nights for 1 to 2 weeks. Then, over the following week, gradually drop to 48° F (9° C) nights. Calceolarias require an average of 4 to 6 weeks of 55° to 58° F (13° to 14° C) night temperatures to develop flower buds. Once buds form, temperatures raised or lowered by 5° F (3° C) will speed up or delay flowering.

Calceolarias are photoperiodic. Be careful to follow the correct culture since calceolarias' response to photoperiod varies by strain. After the cooling period to develop the bud set, give calceolarias such as Brite 'n Early and Glorious Mixture 2 weeks of incandescent lighting from 10 p.m. to 2 a.m. using the same lighting set-up as for chrysanthe-mums. Anytime Mixture and similar varieties are day-length neutral and don't need an extended light period to flower.

| Green Packs | Flower Packs | Crop Time (Weeks) | | No. Plants | | Pinching | Growth Regulators | Garden Height | Staking | Location | Tender/ Hardy |
		Pots 4-in./10-cm.	Baskets 10-in./25.5-cm.	Pot 4-in./10-cm.	Basket 10-in./25.5-cm.						
—	In A-18s, 19-21	19–22	21-23	1	5	No	8	9–12 in. 23–30.5 cm.	***	P. Shade	T

plant in the garden in full to partial shade, spaced 10 in. (25.5 cm.) apart.

Recommended varieties: Nonstops still set the standard for exceptional tuberous-rooted begonias. Offering 10 separate colors and a mixture, the Nonstop series has 2 to 2¹/₂-in. (5 to 6-cm.) blooms that look sharp in borders, baskets and 4-in. (10-cm.) pots.

Other favorites include the Musical series. This pendulous type offers 2- to 2¹/₂-in. double blooms available in five separate colors plus a mixture. Excellent for baskets. Also consider Midnight Beautys, a series of dark-leaved tuberous begonias available in 5 separate flower colors.

*Most often sold by seed count rather than by weight.

**75-78° F will speed up germination but only when used with an overhead mist system. If no mist is used, use lower temperatures of 70-72° F.

***Plants may require staking by late summer, especially if grown in shade.

Green Packs	Flower Packs	Crop Time (Weeks)		No. Plants		Pinching	Growth Regulators	Garden Height	Staking	Location	Tender/ Hardy
7–8	10–12	12–14	14–16	1–2	5–6	No	—	10–12 in.	No	F. Sun	T

container plant or in borders. Plants grow to no more than 10 in. tall in the garden.

Green Packs	Flower Packs	Crop Time (Weeks)		No. Plants		Pinching	Growth Regulators	Garden Height	Staking	Location	Tender/ Hardy
—	14	15–16	20–22	1–2	5–7	No	5, 7	12–18 in. 30.5–46 cm.	No	P. Shade	T

Green Packs	Flower Packs	Crop Time (Weeks)		No. Plants		Pinching	Growth Regulators	Garden Height	Staking	Location	Tender/ Hardy
—	—	22–24	—	1	—	No	6	—	—	—	T

Recommended varieties: Ideal for pot production, Brite 'n Early and Glorious Mixture grow to 9 in. (23 cm.) tall, with blooms up to 1¹/₄ in. (3 cm.) across. Anytime Mixture can be flowered any time during the year in only 18 weeks, without requiring long days and cooler temperatures.

However, with all of our varieties, the highest quality plants are sown in late summer for winter flowering.

Class	Type	Seed for 1,000 plants (oz.)	Number of Seeds	Germination Temperature		Lighting	Days To Germinate	Days Sowing To Transplant	Growing On Temperature	
				Fahrenheit	Celsius				Fahrenheit	Celsius
CALENDULA C. officinalis	A	1/2	3,000/oz. 105/g.	70°	21°	C	10–14	10–15	55°–58°	13°–14°

Ball Culture: Keep temperatures uniform during germination. Calendulas are excellent cool-weather plants that tolerate light frosts, though the flowers may burn. Spaced 12 in. (30.5 cm.) apart, they fill in to create a uniform display in landscape plantings. Plant from early to mid-April in the Midwest for blooming calendulas until July; plants fall apart from the heat in late summer.

In the southern U.S.: Allow 7 weeks for selling dwarf varieties green in packs, 10 weeks for selling them in flower. Plant in full sun, spaced 10 to 12 in. (25.5 to 30.5 cm.) apart. Calendulas do well planted from September to April, though their performance isn't reliable in December and January. These plants rarely survive the heat of summer in the South.

Recommended varieties: The culture information above is based on dwarf series such as Bon Bon or Fiesta Gitana. Best as border plants, both series are excellent choices for 4-in. (10-cm.) pot and pack production, though Fiesta Gitanas are slightly more vigorous.

Class	Type	Seed for 1,000 plants (oz.)	Number of Seeds	Germination Temperature		Lighting	Days To Germinate	Days Sowing To Transplant	Growing On Temperature	
CAMPANULA C. isophylla *Also see Perennial Plants section.*	A, Pe	*	1,500,000– 1,700,000/oz. 52,500–59,500/g.	60°–65°	15°–18°	L	21–28	40–55	60°	15°

Ball Culture: Campanula isophylla is daylength-sensitive; flower buds develop under long days. In our trials, sowings were made in mid-November and grown on under HID lights at 58° to 60° F (14° to 15° C) night temperatures until mid-February. Plants flowered in 4-in. (10-cm.) pots in early April. Without supplemental lighting, the crop time

takes the full 24 weeks to flower from seed when sown November to mid January.

*Most often sold by seed count rather than by weight.

**With the long crop time, these campanulas perform best in pots or baskets.

***Treat as an annual or pot plant in the Midwest. Campanulas are also known to be hardy in the far West and areas of the deep South.

Class	Type	Seed for 1,000 plants (oz.)	Number of Seeds	Germination Temperature		Lighting	Days To Germinate	Days Sowing To Transplant	Growing On Temperature	
CANNA Canna x generalis	A	15	106/oz. 4/g.	70°–75°	21°–24°	C	8–12	15–21	65°	18°

Ball Culture: Cannas from seed have been around for some time. While the most common method of propagation is rhizome multiplication, cannas from seed offer the greenhouse grower an opportunity to produce plants in containers smaller than 6 in. and with a lower initial cost.

Crop time based on 18 cells per flat.

Sowings made in late March and transplanted to A-18's by the middle of April will be salable green in late May. These plants will start flowering by the end of the first week in July. Keep in mind that these cannas will develop a central stem with no basal branching until the primary stem is established. At this time the main stem may be 6-8 in. tall and then dual shoots will emerge at the base of the plant; one on either side of the main stem. These will then

develop and help to fill out the plant. By using a small container, the formation of these secondary stems is delayed due to limited root space; the entire plant's later performance will be spoiled if the plants become rootbound.

If you are using 4-in. pots or larger, the plants will require an additional 1-2 weeks to be salable. In general, one plant per 5-in. pot will flower in 15-16 weeks when grown at

Class	Type	Seed for 1,000 plants (oz.)	Number of Seeds	Germination Temperature		Lighting	Days To Germinate	Days Sowing To Transplant	Growing On Temperature	
CARNATIONS Dianthus caryophyllus *Also see Perennial Plants section.*	A	1/8	14,000/oz. 490/g.	65°–70°	18°–21°	C. Lt.	5–13	15–20	50°–55°	10°–13°

Ball Culture: Chill seed for 1 to 2 weeks at 30° to 35° F (-1° to 2° C) before sowing to aid germination. Keep germination temperatures uniform. Allow 17 to 19 weeks in the southern and far western states for flowering 4-in. (10-cm.) pots. In the Midwest, an early December sowing will flower

in late May or June, while an early September sowing will flower in early to mid-February. It's imperative to maintain high light quality for winter flowering crops, which require additional lighting in the northern states.

In the southern U.S.: Allow 17 to 19 weeks for flowering pots in the southern and far western states; allow 11 weeks for selling green in the pack in the deep South. Carnations can be planted from August to January for blooming plants until June, though they are not noted for their free-flower

BEDDING, FLORIST & FOLIAGE PLANTS

Crop Time (Weeks)				No. Plants							
Green Packs	Flower Packs	Pots 4-in./10-cm.	Baskets 10-in./25.5-cm.	Pot 4-in./10-cm.	Basket 10-in./25.5-cm.	Pinching	Growth Regulators	Garden Height	Staking	Location	Tender/ Hardy
8–10	12–14	13–15	—	1	—	No	7	*	*	F. Sun	H

*Height and staking requirements are determined by type and garden use. With varieties ranging from dwarf to tall, calendulas stand anywhere from 10 to 28 in. (25.5 to 68.5 cm.) high. As cut flowers, the taller types require 1 layer of support to keep them upright and prevent stems from bending.

In outdoor plantings, however, taller calendulas are past their prime by the time they become top-heavy, so staking isn't needed.

Green Packs	Flower Packs	Pots 4-in./10-cm.	Baskets 10-in./25.5-cm.	Pot 4-in./10-cm.	Basket 10-in./25.5-cm.	Pinching	Growth Regulators	Garden Height	Staking	Location	Tender/ Hardy
**	**	23-26	25-27	1	5	No	—	8 in. 20 cm.	No	F. Sun to P. Shade	***

Green Packs	Flower Packs	Pots 4-in./10-cm.	Baskets 10-in./25.5-cm.	Pot 4-in./10-cm.	Basket 10-in./25.5-cm.	Pinching	Growth Regulators	Garden Height	Staking	Location	Tender/ Hardy
9–11	*	*	—	*	—	No	—	2.5–3 ft.	No	F. Sun	T

60-65° F. If using lower temperatures, add additonal time. Too cool (below 54° F) will stall the plant and can damage it.

In the southern U.S., cannas are often grown for sales in winter for color during the spring and summer. In the warmer areas of the deep South, the plants are left in the ground and divided about every third year. In areas farther north where winters are cold, the rhizomes are dug and

stored in the home or garage for the winter.

Recommended varieties: The culture noted above is based on the All America Award Winner Tropical Rose. This variety bears 3-4-in. rose-pink flowers on plants which reach no more that 3 ft. in height. Additional work is being done on introducing a red-flowering counterpart, but it has yet to be released at the time of this writing.

*Cannas should not be grown in a container smaller than 18 cells per flat or 4-in. pot. For effective flowering and root development, it is suggested that you use nothing smaller that a 5-in. pot. Green pack sales, however, can be produced in a 32-tray but plants will require up to four weeks to flower after transplant to the garden if grown this way.

Green Packs	Flower Packs	Pots 4-in./10-cm.	Baskets 10-in./25.5-cm.	Pot 4-in./10-cm.	Basket 10-in./25.5-cm.	Pinching	Growth Regulators	Garden Height	Staking	Location	Tender/ Hardy
12–16	*	22–25	—	1–2	—	**	8	8–16 in. 20–41 cm	No	F. Sun to P. Shady	H

ing performance. For best effect in gardens and landscapes, space 10 in. (25.5 cm.) apart in staggered rows. In the deep South, carnations should be planted in partial shade.

Recommended varieties: The Monarch and Lillipot series are outstanding varieties for 4 and 6-in. (10 and 15-cm.) pot production. They also do well in packs and offer the earliest flowering of any F$_1$ carnation from seed, including older varieties like Chaubaud.

*Because of the long crop time, flowering packs are not recommended in the midwestern and eastern states.

**Dwarf varieties (such as the Lillipot and the Monarch series) don't require pinching.

Class	Type	Seed for 1,000 plants (oz.)	Number of Seeds	Germination Temperature		Lighting	Days To Germinate	Days Sowing To Transplant	Growing On Temperature	
				Fahrenheit	Celsius				Fahrenheit	Celsius
CASSIA C. alata (Christmas candle)	F	2	650/oz. 23/g.	Day: 85° Night: 70°	Day: 29° Night: 21°	C	5–8	14°–20°	60°–62°	15°–17°

Ball Culture: Don't presoak seed prior to sowing. Cassias are most often grown in large containers for green foliage sales; they can also be used as bedding plants for background accents. Spring sowings don't flower in the summer, but produce lush, deep green foliage instead. April sowings in the Midwest produce 4 to 5 ft. (1.2 to 1.5 m.) plants by mid to late September if grown warm. Cassias cannot tolerate cold, or even cool weather, and die at the first sign of frost.

In the southern U.S.: Sowings made in January and planted to the garden in March have been known to flower from August until the first frost. When flowering, plants reach 6 to 8 ft. (1.8 to 2.4 m.) tall.

Class	Type	Seed for 1,000 plants (oz.)	Number of Seeds	Germination Temperature		Lighting	Days To Germinate	Days Sowing To Transplant	Growing On Temperature	
				Fahrenheit	Celsius				Fahrenheit	Celsius
CELOSIA C. plumosus (feather or plume)	A,C	1/32–1/16	39,000/oz. 1,365/g.	75°	24°	C	8–10	10–15	65°–68°	18°–20°
C. cristata (crested)	A,C	1/16	34,000/oz. 1,190/g.	75°	24°	C	8–10	10–15	65°–68°	18°–20°
C. cristata forma spicata	A	1/16	39,000/oz 1,375/g.	75°	24°	C	5–10	18–28	65°–68°	18°–20°

Comments: Growing too cool or planting outdoors too early will cause premature flowering and spoil later performance. The crop's wide range of colors makes it ideal for landscaping. Varieties like the Century and Feather series will fill in quickly and flower soon after, spaced 12 in. (30.5 cm) apart.

In the southern U.S.: Allow 7 to 9 weeks for green pack sales, and 12 to 13 weeks for 4-in. (10 cm.) flowering pots.

Plant after all danger of frost has passed, until early April. If established before the weather gets hot, celcsias will flower until August or September. Plant in full sun, with dwarf varieties spaced 8 to 10 in. (20 to 25.5 cm.) apart and taller varieties 12 to 18 in. (30.5 to 46 cm.) apart. Among the dwarf varieties, the feather or ploom types show good color through mid-July, while crested types continue to display well until August. The taller varieties put on a good performance until the first frost.

Recommended varieties: Choose varieties carefully for specific uses. The Century series and their upright, spreading celosias as tall as 20 in. (51 cm.) are ideal for mass plantings, filling in well on 12 in. (30.5 cm.) centers. The Centurys hold up longer and their colors fade less than other varieties, placing them among our top recommendations for overall performance. Smaller, dwarf varieties, such as Geisha, Castle, and Fairy Fountains make excellent border and 4-in. (10-cm.) pot plants, though the flowers aren't as brilliant.

Class	Type	Seed for 1,000 plants (oz.)	Number of Seeds	Germination Temperature		Lighting	Days To Germinate	Days Sowing To Transplant	Growing On Temperature	
				Fahrenheit	Celsius				Fahrenheit	Celsius
CHRISTMAS CHERRY Solanum pseudocapsicum	P	1/8	12,000/oz. 420/g.	70°	21°	L	7-15	21-28	55°	13°

Ball Culture: Sow from February 15 to March 1 for early December sales. Keep soil temperatures at 70° F (21° C) or above for best germination. Don't cover the seed, but expose it alternately to light and dark. If using lights in the germination area of the greenhouse or growth chamber, turn those lights out each night; germination percentages may increase using this method. Upon germination, drop night temperatures to 60° to 62° F (15° to 17° C) to establish plants in packs or small pots (2 1/4 in. / 5.6 cm.).

Christmas cherries are slow growers that need to be transplanted as the roots fill containers. Never let plants become root-bound unless they are in the final pots. Throughout the summer, grow plants in 6-in. (15-in.) pots outdoors where the wind will pollinate flowers and

Green Packs	Crop Time (Weeks) Flower Packs	Pots 4-in./10-cm.	Baskets 10-in./25.5-cm.	No. Plants Pot 4-in./10-cm.	Basket 10-in./25.5-cm.	Pinching	Growth Regulators	Garden Height	Staking	Location	Tender/ Hardy
7–9	—	11–13	—	1	—	No	—	3–5 ft. .9–1.5 m.	No	F. Sun	T
7–9	10-12	13	—	1	—	No	1	10-36 in. 25.5–91.5 cm.	No	F. Sun	T
7–9	10-12	13	—	1	—	No	1	5–32 in. 13–81 cm.	*	F. Sun	T
7–9	9–10	11-14	—	1	—	No	No	18–26 in. 46-66 cm.	No	F. Sun	T

Among the crested types, try the Jewel Box series or Olympia Mixture for 4-in. (10-cm.) pot and pack sales as border plants.

For cut flowers, select taller varieties that hold their color well once cut. Among the feather types, the Century series and our red-flowered Forest Fire are excellent for medium length cuts. Crested cut types also make good cut flowers, such as Toreador, a dark red variety, and the Chiefs, an upright series

reaching 3 ft. (91.5 cm.) tall.

C. crista forma spicatra: Characterized by only one variety at the time of this writing, Flamingo Feather is a predominantly upright growing plant with light rose pink flowers which fade slightly as they age under the sun. To use celosia as a cut flower, harvest the flower heads realizing that the color will lighten readily.

Sowings made in early April will flower by the first week of June when grown one plant per cell in a 32-cell pack with 55° to 60°F night temperatures, and daytime temperatures 5 to 10° warmer. Plants grow erect with limited basal branching until flowering begins. Once flowering starts, plants will branch more readily but will only spread to 6 in. across.

*Staking is required for taller, upright varieties, especially crested types.

Green Packs	Flower Packs	Pots 4-in./10-cm.	Baskets 10-in./25.5-cm.	Pot 4-in./10-cm.	Basket 10-in./25.5-cm.	Pinching	Growth Regulators	Garden Height	Staking	Location	Tender/ Hardy
—	—	6 in./15 cm. pot 40–50	—	6 in./15 cm. pot 1–3	—	Yes*	—	8–13 in. 20–33 cm.	No	F. Sun	T

encourage fruit set. A cold frame with raised sides can be used for maximum air circulation. Plants grown outside in frames should be brought back into the greenhouse in September. Grow on at 55° F (13° C) nights until sold—plants can be held at temperatures of 50° to 55° F (10° to 13° C) for 2 weeks to tone them up before Christmas sales.

*Christmas cherries should be pinched several times to encourage branching, but don't pinch plants after July 1.

Class	Type	Seed for 1,000 plants (oz.)	Number of Seeds	Germination Temperature		Lighting	Days To Germinate	Days Sowing To Transplant	Growing On Temperature	
				Fahrenheit	Celsius				Fahrenheit	Celsius
CHRYSANTHEMUM, Button Types C. multicaule (Yellow Buttons)	A	1/16–1/8	20,000/oz. 700/g.	60°–65°	15°–18°	C. Lt.	10–14	11–14	60°–65°	15°–18°
C. paludosum (White Buttons)	A	1/32	50,000/oz. 1,750/g.	60°–65°	15°–18°	C. Lt.	10–14	11–14	60°–65°	15°–18°

Ball Culture: These chrysanthemums grow and bloom quickly. Yellow Buttons flowers 3 to 7 days earlier than White Buttons.

In our Chicago-area trial, this crop performs well right into August, until the heat 'melts' them. Used as border plants in locations with afternoon shade, chrysanthemums will continue to provide color until frost. We don't recommend them as indoor pot plants since the foliage has a strong scent, especially C. paludosum.

Class	Type	Seed for 1,000 plants (oz.)	Number of Seeds	Germination Temperature		Lighting	Days To Germinate	Days Sowing To Transplant	Growing On Temperature	
CINERARIA Senecio cruentus	P	1/128–1/64	150,000/oz. 5,250/g.	70°–75°	21°–24°	L	10–14	18–24	60° *	15° *

Ball Culture: When established in the final container, 1 plant per 4- to 6-in. (10 or 15-cm.) pot, drop night temperatures to 50° to 55° F (10° to 13° C), encouraging uniform bud set and compact habit development. Flower buds will be visible in 3 to 4 weeks; add several weeks to this time when growing temperatures are 5° to 10° F (3° to 6° C) higher. Once visible, the buds will take another 7 to 9 weeks to bloom at 55° F (13° C) nights. Flowering can be delayed by growing on at night temperatures below 50° F (10° C), or hastened by increasing the night temperature to 60° F (15° C). However, more vegetative growth should be expected at 60° F (15° C).

This crop will not bloom again once the first flowers die out. As pot plants, cinerarias will stay in bloom 2 to 3

Class	Type	Seed for 1,000 plants (oz.)	Number of Seeds	Germination Temperature		Lighting	Days To Germinate	Days Sowing To Transplant	Growing On Temperature	
CLEOME C. hasslerana	A, C	1/8	14,000/oz. 490/g.	Day: 80° Night: 70°	Day: 26° Night: 21°	C. Lt.	10–12	21–25	70°–75°	21°–24°

Ball Culture: Cleomes have strongly scented foliage that may be overpowering in small, enclosed gardens. Plants also have short, sharp spurs similar to thorns along the stems.

For best results, alternate germination temperatures. Cleomes can be sold green in 4-in. (10-cm.) pots since the plants may be rather tall by the time they flower. After 7 to 9 weeks, plants will already be 5 to 8 in. (13 to 20 cm.) tall in packs.

To disguise cleomes' open growth habit, the plants should be spaced 15 to 18 in. (38 to 46 cm.) apart in displays. Plants begin branching in the garden when stems get about 12 to 14 in. (30.5 to 35.5 cm.) tall, and spread up to 40 in.

Class	Type	Seed for 1,000 plants (oz.)	Number of Seeds	Germination Temperature		Lighting	Days To Germinate	Days Sowing To Transplant	Growing On Temperature	
COBAEA C. scandens (Cup and Saucer Vine, Cathedral Bells)	A	5	375/oz. 13/g.	70°	21°	L/C	4–8	15–20	50°–55°	10°–13°

Comments: This vigorous climber is native to Mexico. Plants grow best where they get full sun to partial shade and are allowed to grab hold of a wall or facade. Blooms are to 2½ inches across and are purple blue in color. Flowers have no scent but are plentiful when the seed is sown early in the season. Seed can be sown direct to the final container or plug tray to avoid transplant shock, though individual seedling transplant can be done with limited damage or problems.

In general, sowings made in April will produce seedlings large enough to transplant to cell packs or pots within 20 days if the night temperatures are in the 60's or upper 50's. While we have grown these plants with success in cell packs, their best performance comes from a container and trellis (12- to 15-in. high) system that is often used on clematis. However, Cobaea is much more vigorous. Older references suggest that a March sowing will produce blooming plants mid summer until fall. Our trials indicate that an April sowing will produce blooming plants by Labor Day and these plants, though fully leaved out, seldom reach flowering potential before frost.

BEDDING, FLORIST & FOLIAGE PLANTS

| | | Crop Time (Weeks) | | No. Plants | | | | | | | |
| | | Pots | Baskets | Pot | Basket | | Growth | Garden | | | Tender/ |
Green Packs	Flower Packs	4-in./10-cm.	10-in./25.5-cm.	4-in./10-cm.	10-in./25.5-cm.	Pinching	Regulators	Height	Staking	Location	Hardy
—	10–11	11–12	—	1–2	—	No	2	8–12 in. 20–30.5 cm.	No	F. Sun	T
—	10–11	11–12	—	1–2	—	No	2	8–12 in. 20–30.5 cm.	No	F. Sun	T
—	—	6 in./15 cm. pots 27	—	1	—	No	6	—	No	—	T
7–9	—	—	—	1	—	No*	1	3–5 ft. .9–1.5 m.	No**	F. Sun	T
8–10	—	—	—	1	5	Yes	No	Vine	Yes	F. Sun to P. Shade	T

weeks when kept in bright areas and provided with ample water. As bedding plants, they perform best in cool regions like coastal California. Planted to the bed when just beginning to show color, but before full bloom, this annual will remain in flower 1 to 2 months.

*Drop night temperatures to 50° to 55° F (10° to 13° C) once established in final containers.

(1 m.). Cleomes fill in quickly and flower until frost.

In the southern U.S.: Plant from summer or early fall until late December. Cleomes have not performed very successfully in the deep South, including Florida and southern Texas.

* Pinching results in top-heavy plants.

**Staking is only required late in the summer if plants are grown out in the open instead of against a wall or fence, or if landscapers or home gardeners pinch them back during the growing season.

In his book, *A Little Book of Climbing Plants* (A.T. De La Mare Co., 1924), Alfred C. Hotte comments that seed was started in February, the plants were moved up into larger containers as the season progressed and were pinched back a number of times to produce 6 to 8 stems to keep the plants busy. The greenhouse temperatures after roots are established in the initial container were about 50° F. Keep in mind that the plants will be large at the time they are sold.

One final note: There are some variegated forms of these plants which should be vegetatively propagated by cuttings when a particular selection piques your interest.

Class	Type	Seed for 1,000 plants (oz.)	Number of Seeds	Germination Temperature Fahrenheit	Celsius	Lighting	Days To Germinate	Days Sowing To Transplant	Growing On Temperature Fahrenheit	Celsius
COFFEE Coffea arabica	F	11	140/oz. 5/g.	76°	24°	C	35	—	65°–75°	18°–24°

Ball Culture: Because germination is variable, soak seed overnight prior to sowing or, better yet, scratch or file the seed coat before submersing in water to increase the germination rate. Coffee plants tend to grow upright with few branches; even pinching does not induce them to branch well. Allow 32 to 36 weeks to sell green in 6-in. (15-cm.) pots, 3 plants per pot.

Class	Type	Seed for 1,000 plants (oz.)	Number of Seeds	Germination Temperature Fahrenheit	Celsius	Lighting	Days To Germinate	Days Sowing To Transplant	Growing On Temperature Fahrenheit	Celsius
COLEUS C. x hybridus	A	1/64	100,000/oz. 3,500/g.	70°–75°	21°–24°	L	10–14	20–25	65°–75°	18°–24°

Ball Culture: Since coleus are sold for their colorful foliage rather than their flowers, varieties offering slow flower development, such as the Wizard series, are most popular. Intensity of foliage color—especially important to landscape and garden designers—is determined by the amount of light provided. For example, in shaded plantings, Wizard Sunset is bright bronze-scarlet, but in sun its leaves turn deep, dark red.

In the southern U.S.: Allow 8 to 9 weeks for pack sales, 11 to 13 for 4-in. (10-cm.) pots. Plant in early spring after all danger of frost has passed. Pinching may be required. In the garden, coleus do best in partial shade.

Recommended varieties: Selections are largely based on leaf shapes and sizes; see the Ball Seed Catalog for more details. Among the large, robust coleuses, we suggest the Rainbow Mix or Wizard Series. The heart-shaped leaves of these 2 series are the largest we offer. Our bushy, compact Carefree series features the smallest leaves, with attractive, deeply curled edges.

Class	Type	Seed for 1,000 plants (oz.)	Number of Seeds	Germination Temperature Fahrenheit	Celsius	Lighting	Days To Germinate	Days Sowing To Transplant	Growing On Temperature Fahrenheit	Celsius
COSMOS C. sulphureus	A	1/2	4,000/oz. 70/g.	70°	21°	C	5–7	11–15	65°	18°
C. bipinnatus	A	1/2	4,000/oz. 70/g.	70°	21°	C	5–7	11–15	65°	18°

Ball Culture: Even after just 5 weeks, either type of cosmos can be 6 to 12 in. (15 to 30.5 cm.) tall in the pack without growth regulators. For flowering pot sales of sulphureus varieties in the spring, use 1 plant per 4-in. (10-cm.) pot or 3 per 6-in. (15-cm.) pot. We recommend dwarfs such as the Sunny series, since cosmos tend to be rather weak-stemmed and some staking is needed for the taller types. In landscape plantings, sulphureus varieties are often short-lived, performing well only until August when they become ragged-looking.

The bipinnatus species consists of predominantly short day plants that flower during the late summer and fall when sown in the spring. In spring, sowings made after March 1 should be given 10 hours or less of light to flower cosmos like Imperial Pink and Sea Shells. Newer series like Sonata and the Sonata Mixture variety, do not require short days to

Class	Type	Seed for 1,000 plants (oz.)	Number of Seeds	Germination Temperature Fahrenheit	Celsius	Lighting	Days To Germinate	Days Sowing To Transplant	Growing On Temperature Fahrenheit	Celsius
CRAPE MYRTLE Lagerstroemia indica	A	1/8	12,000/oz. 420/g.	70°	21°	C	14	20–25	65°–68°	18°–20°

Ball Culture: Considered a pot plant in the northern states, crape myrtles are used as perennial shrubs in the South. Allow 36 to 40 weeks for growing in 4 or 5½-in. (10 or 14-cm.) pots.

Crape myrtles will not come into uniform bloom from seed when grown in packs, so use in pot production only.

Crop Time (Weeks)				No. Plants								
Green Packs	Flower Packs	Pots 4-in./10-cm.	Baskets 10-in./25.5-cm.	Pot 4-in./10-cm.	Basket 10-in./25.5-cm.	Pinching	Growth Regulators	Garden Height	Staking	Location	Tender/ Hardy
—	—	28	—	1	—	No	—	—	—	—	T
9–10*	—	12–13*	14–16*	1	5–6	Yes**	1, 3	10–16 in. 25–41 cm.	No	Shade to P. Sun	T

*The culture above is based on the Wizard series; crop times for other coleus, such as the Saber, Carefree and Fiji Mixtures, should be increased by 1 to 2 weeks.

**Older varieties of coleus require pinching of flower buds to prevent flower development. Wizards are later to flower and don't normally require pinching. However, in areas with long growing seasons, plants may need to be pinched in the garden to stop flowering. The latest varieties in the Wizard series to flower include, Scarlet, Golden and Sunset.

4–6	—	—	—	1	—	No	2	16–24 in. 41–61 cm.	*	F. Sun	T
4–6	—	—	—	—	—	No	2	36–48 in. 91.5–122 cm.	*	F. Sun	T

bloom. Collectively, following the procedures above, both neutral and short day plants will flower in June from a March sowing. Be aware that any bipinnatus varieties grown in cell packs will be large and robust when flowering and will perform better if sold green or with early bud.

In the southern U.S.: Once all danger of frost has passed, plant in full sun until April for flowering throughout the summer. Cosmos have not performed well in the deep South, where plants can reach up to 5 ft. (1.5 m.) in height.

*Taller varieties may require staking.

—	—	36–40	—	1	—	No	—	—	No	F. Sun	—

BEDDING, FLORIST & FOLIAGE PLANTS

Class	Type	Seed for 1,000 plants (oz.)	Number of Seeds	Germination Temperature Fahrenheit	Celsius	Lighting	Days To Germinate	Days Sowing To Transplant	Growing On Temperature Fahrenheit	Celsius
CROSSANDRA C. infundibuliformis	F	1/2	4,000/oz. 140/g.	Day: 80° Night: 70°	Day: 26° Night: 21°	C	21–28	—	65°–68°	18°–20°
CUPHEA C. platycentra	A	1/8	16,000/oz. 560/g.	70°	21°	C. Lt.	12–15	21–25	65°	18°
CYCLAMEN C. persicum	P	1/2*	2,500/oz. 87/g.	60°	15°	C	28–35	50–70	—	—
CYNOGLOSSUM C. amabile (Chinese Forget-Me-Not)	A	1/4 to 1/2	5,000/oz. 176/g.	65°–70°	18°–21°	C	5–8	14–21	60°	15°
DAHLBERG DAISY Dyssodia tenuiloba	A	1/64	180,000/oz. 6,300/g.	65°–70°	18°–21°	L	10–16	20–26	65°–68°	18°–20°

CROSSANDRA

Ball Culture: Germination usually starts 10 to 14 days after sowing, and continues slowly and irregularly over a period of a month. Transplanting, therefore, must be done at intervals as plants appear and become large enough to handle.

Sell as foliage or pot plants in 4- and 6-in. (10- and 15-cm.) pots. In northern climates, crossandras will bloom in 7 months for summer flowering, and in 9 months for winter flowering.

CUPHEA

Ball Culture: Cuphea is a relatively easy crop to grow from seed; however, germination temperatures are critical. In landscapes, the plants appear bronze when planted in full sun and reach no more than 8 to 10 in. (20 to 25.5 cm.) tall. In shady areas, cupheas grow as high as 12 in. (30.5 cm.).

In the southern U.S.: Allow 10 to 12 weeks for flowering pack sales.

CYCLAMEN

Ball Culture: Pour warm water (70° to 80° F/21° to 26° C) over seed and let it soak overnight. After sowing 1/4 to 1/2 in. (.6 to 1 cm.) deep, sprinkle a layer of shredded sphagnum peat moss over covered seed to help maintain uniform moisture. Keep flats in the dark at a constant temperature of 60° F (15° C). After germination, drop the temperature to 55° F (13° C). As the seedlings begin to crowd, transplant into 3-in. (7.5-cm.) pots or flats, spacing at 3 x 3 in. (7.5 x 7.5-cm.).

Allow 9 to 11 months for flowering 6-in. (15-cm.) pots. Keep in mind that cyclamens need shade and ventilation during the summer months. For Christmas flowering, make the final shift into 4, 5 or 6-in. (10, 13 or 15-cm.) pots in August or early September, and grow on at 50° F (10° C). Cyclamens are a popular choice for pots, beds and baskets in the fall or winter. This perennial is also used occasionally

CYNOGLOSSUM

An old-time annual that could use some reselection for longevity in the garden. Single flowers are light to medium blue on plants that reach no more than 20 in. in the garden. Sowings made in early March and planted to the garden by the end of May will begin flowering by the end of June. Plants will reach their peak flower power by the middle of July and often finish blooming between the middle and end of August.

It appears that a light trimming after flowering encourages the plant to bloom again around the end of August. However, this is not foolproof and often the plants are dead by the middle of September.

DAHLBERG DAISY

Ball Culture: Germination on this crop is irregular and seed can go dormant, so order seed accordingly. Dahlberg daisies work well in hanging baskets and pots and look great as bedding plants in the landscape. Plants tend to fall over readily, which is ideal for baskets but not for packs.

Plants can, however, be sold green or with budded color in packs.

To create a bright mass of color in the garden, space 8 to 10 in. (20 to 25.5 cm.) apart.

BEDDING, FLORIST & FOLIAGE PLANTS

Green Packs	Flower Packs	Crop Time (Weeks) Pots 4-in./10-cm.	Baskets 10-in./25.5-cm.	No. Plants Pot 4-in./10-cm.	Basket 10-in./25.5-cm.	Pinching	Growth Regulators	Garden Height	Staking	Location	Tender/ Hardy
—	—	30–34	—	1	—	No	—	—	—	—	T
9–10	11–12	13–15	—	1	—	No	—	8–10 in. 20–25.5 cm	No	F. Sun to P. Shade	T
—	—	—	—	—	—	No	—	—	—	—	T
9–10	—	—	—	—	—	No	No	18–20 in. 46–51 cm.	No	F. Sun	T
8–9	10–11	11–12	12–13	1–3	5–7	No	—	6–8 in. 15–20 cm.	No	F. Sun	T

as a bedding plant in cool coastal climates, as in California.

Recommended varieties: For earlier, more uniform and longer-lasting blooms, F$_1$ hybrids such as the Sierra series

and the Pannevis line, now called Concerto, are by far the most popular choices. The Sierras are ideal for 6-in. (15-cm.) pot production, blooming uniformly throughout the color range. The Concerto line is made up of many favorites, including Esmeralda, Finlandia and Norma.

Offering the broadest selection of colors, the Concerto series is noted for both its free-flowering performance and large flower size.

*Most often sold by seed count rather than by weight.

BEDDING, FLORIST & FOLIAGE PLANTS *(vertical side tab)*

Class	Type	Seed for 1,000 plants (oz.)	Number of Seeds	Germination Temperature Fahrenheit	Celsius	Lighting	Days To Germinate	Days Sowing To Transplant	Growing On Temperature Fahrenheit	Celsius
DAHLIA D. x hybrida	A	1/2	2,800/oz. 98/g.	60°–65°	15°–18°	C	5–10	11–15	55°–60°	13°–15°

Ball Culture: Keep soil temperature constant during germination. Some colors in a mixture may produce weaker seedlings than others, but when most seedlings are ready, transplant them all. The weaker seedlings often develop into strong plants. Although seed dahlias make excellent bedding, border and pot plants, they don't flower as pro-fusely as other bedding plants and are best used sparsely in landscaping. Space 12 in. (30.5 cm.) apart in gardens and landscape plantings. For Mother's Day sales in 4-in. (10-cm.) pots, sow in early February. If amply spaced and fed, no pinching is needed.

In the southern U.S.: Allow 9 to 10 weeks for flowering packs and 12 weeks for flowering 4-in. (10-cm.) pots, using 1 plant per pot. For best performance, plant after all danger of frost has passed until early May. Dahlias continue to bloom up to September; however, flowers appear sporadically, at best, from July on. Plant dahlias in full sun or

Class	Type	Seed for 1,000 plants (oz.)	Number of Seeds	Germination Temperature Fahrenheit	Celsius	Lighting	Days To Germinate	Days Sowing To Transplant	Growing On Temperature Fahrenheit	Celsius
DIANTHUS D. chinensis	A	1/16	25,000/oz. 875/g.	70°–75°	21°–24°	C. Lt.	7	18–25	50°–55°	10°–13°

Ball Culture: At the Chicago latitude, the Princess and Telstar series have delivered excellent landscape performances. When planted to the open ground showing budded color while the soil is still cool (around mid-May), these two series flower all summer long. Plants can be spaced 10 in. (25.5 cm.) apart to fill in, or 12 in. (30.5 cm.) apart for some room between them. Though they remain in flower, dianthus may heat stall in August for several weeks.

Dianthus are daylight-sensitive, so for Valentine's Day and other winter sales, plants need bright lighting or days extended by at least 4 hours to produce uniform plants.

In the southern U.S.: Sow seed in August for green pack sales in November. Plants will flower from February until May or possibly June. Allow 9 to 10 weeks for green packs and 11 to 12 weeks for flowering 4-in. (10-cm.) pots, using 1 plant per pot. Plants can be spaced at 10 or 12 in. (25.5 or 30.5 cm.) in the garden, depending on the effect desired.

Class	Type	Seed for 1,000 plants (oz.)	Number of Seeds	Germination Temperature Fahrenheit	Celsius	Lighting	Days To Germinate	Days Sowing To Transplant	Growing On Temperature Fahrenheit	Celsius
DOLICHOS D. lablab	A	14	110/oz. 4/g.	70°–72°	21°–22°	C	5–8	*	60°	15°

Comments: Though not popular in the United States, Dolichos is well known as a vegetable plant throughout various regions around the world. Its purple-hued stems, leaves, and seed pods are complimented by the lavender-bluish flowers produced by this vining plant. In the U.S., use Dolichos to cover trellises or as a backdrop for garden walls.

*Sow 2-3 seeds per Jiffy pot rather than attempting single-seedling transplant, which can weaken and kill the seedlings. Instead, once the seedlings emerge in the Jiffy pot, allow 2-4 weeks for salable green plants and then tell your customers to remove the bottom of the pots as they plant them in their garden. If allowed to remain in the pot beyond 5-7 weeks before planting, they are often stunted in the garden and never attain their desired height.

Class	Type	Seed for 1,000 plants (oz.)	Number of Seeds	Germination Temperature Fahrenheit	Celsius	Lighting	Days To Germinate	Days Sowing To Transplant	Growing On Temperature Fahrenheit	Celsius
DRACAENA Cordyline indivisa	F, A	1/4	10,000/oz. 350/g.	Day: 86° Night: 68°	Day: 30° Night: 20°	L	30–40	60–95	65°–68°	18°–20°

Ball Culture: Dracaena is a foliage item most commonly treated as a bedding plant. Sow in August for May sales of green 3-in. (7.5-cm.) pots, or in December for 2½-in. (6-cm.) pot sales. Allow 15 months for strong 4 to 5-in. (10 to 13-cm.) plants, sowing from February to March for sales the following spring.

Dracaenas can be planted in landscapes in full sun, or in containers with other annuals in partial shade. Spaced 10 to 12 in. (25.5 to 30.5 cm.) apart, try a double row of dracaenas in the center of a bed of Red Bandit and Pinto Salmon geraniums for a dramatic impact.

BEDDING, FLORIST & FOLIAGE PLANTS

Crop Time (Weeks)				No. Plants		Pinching	Growth Regulators	Garden Height	Staking	Location	Tender/ Hardy
Green Packs	Flower Packs	Pots 4-in./10-cm.	Baskets 10-in./25.5-cm.	Pot 4-in./10-cm.	Basket 10-in./25.5-cm.						
10–11	12–13	13–14	—	1	—	No	2, 5, 8	8–27 in. 20–68.5 cm.	No	F. Sun to P. Shade	T

afternoon shade, space 12 in. (30.5 cm.) apart.

As cut flowers, dahlia stems should be cut and seared to prevent wilting and loss of sap.

10–11	15–16*	15–16	—	1	—	No	1, 3, 8	9–12 in. 23–30.5 cm.	No	F. Sun to P. Shade	H

Recommended varieties: All of our dwarf dianthus make excellent 4-in. (10-cm.) pot plants. However, our earliest and most heat-tolerant series, the Telstar and Princess, offer the best performance in packs, pots and landscape plantings.

*Even under cooler conditions dianthus have a short shelf life, so overall quality is best when plants are sold green or with early color.

—	—	—	—	—	—	No	—	3–6 ft. 9–1.8 m.	Yes	F. Sun to P. Shade	T

—	—	—	—	1	—	No	1	18–24 in. 46–61 cm.	No	F. Sun to P. Shade	T

BEDDING, FLORIST & FOLIAGE PLANTS *(side tab)*

Class	Type	Seed for 1,000 plants (oz.)	Number of Seeds	Germination Temperature Fahrenheit	Celsius	Lighting	Days To Germinate	Days Sowing To Transplant	Growing On Temperature Fahrenheit	Celsius
DUSTY MILLER Senecio cineraria (Silverdust)	A, C	1/32	50,000–100,000/oz. 1,750–3,500/g.	72°–75°	22°–24°	L	10–15	20–25	60°–65°	15°–18°
Cineraria maritima (Diamond)	A, C	1/32	50,000/oz. 1,750/g.	72°–75°	22°–24°	L	10–15	20–25	60°–65°	15°–18°
Chrysanthemum ptarmiciflorum (Silver Lace)	A, C	1/128	200,000/oz. 7,000/g.	72°–75°	22°–24°	L	10–15	20–25	65°	18°

Ball Culture: Noted for their silvery leaves, dusty millers are considered tender biennials in the northern U.S., though they are officially listed as annuals. These foliage varieties are excellent for container, bedding and landscape plantings, as well as cut and dried for floral arrangements. Lines differ mostly by leaf size and shape, though there are other distinguishing features. Silverdust has the most silvery foliage of our dusty millers. Standing just 6 to 8 in. (15 to 20 cm.) tall, Silver Lace is our smallest dwarf. Both can be spaced 10 in. (25.5 cm.) apart in beds, but Silverdust fills in more quickly than Silver Lace. Another popular line, Maritima Diamond, is a more upright and vig- orous version of Silverdust, with a similar leaf formation. On 12- in. (30.5- cm.) centers, Diamond fills in just 4 to 5 weeks after planting to the garden.

Dusty millers, particularly Silverdust, are very hardy and can usually survive fall temperatures in the Midwest.

Class	Type	Seed for 1,000 plants (oz.)	Number of Seeds	Germination Temperature Fahrenheit	Celsius	Lighting	Days To Germinate	Days Sowing To Transplant	Growing On Temperature Fahrenheit	Celsius
ESCHSCHOLZIA E. californica (California Poppy)	A*	1/8	17,000/oz. 600/g.	70°–72°	21°–22°	—	4–8	**	60°	15°

Ball Culture: Several seeds should be sown per cell in packs, allowing for 2-3 finished plants per cell. Sowings made in early April will be salable green by the end of the month of May, though the lower foliage is often yellowing by this time. The stress created by cool temperatures, mini- mal watering, and low fertility often creates this problem.

However, growing warmer (above 62° F nights) often leads to stretched and weak growth. Flowers will close by 6 or 7 pm and will reopen the following morning.

*California poppies are classified as short-lived perennials and are treated in many areas as reseeding annuals.

**Sow seed direct to the final container. Plants resent transplanting and individually transplanted seedlings often stall and die soon after transplanting is attempted.

In the southern U.S.: Most often sown direct to the flower bed in the fall of the year, California poppies will flower

EUSTOMA
E. grandiflorum (formerly E. russellianum)
See Lisianthus, Cut Flowers section.

Class	Type	Seed for 1,000 plants (oz.)	Number of Seeds	Germination Temperature Fahrenheit	Celsius	Lighting	Days To Germinate	Days Sowing To Transplant	Growing On Temperature Fahrenheit	Celsius
EXACUM E. affine (Persian violet)	P	1/128 = 5,468 seeds	1,000,000/oz. 35,000/g.	70°	21°	C	14–21	35–42	65°–68°	18°–20°

Ball Culture: Exacum is a fragrant greenhouse pot plant that can be grown year-round, though it is not recommend- ed as a winter crop in the North. Once sown, cover seed lightly and mist rather than water in the flat. Transplant seedlings into 2½-in. (6-cm.) pots or packs when they begin to crowd. After moving to the final pot, grow on at 65° F (18° C), and shade from the summer sun. Allow 4 to 5 months for flowering 5-in. (13-cm.) pots, with 1 to 3 plants per pot.

*For 5-in. (13-cm.) pots, use 1 to 3 plants each.

BEDDING, FLORIST & FOLIAGE PLANTS

Crop Time (Weeks)				No. Plants		Pinching	Growth Regulators	Garden Height	Staking	Location	Tender/ Hardy
Green Packs	Flower Packs	Pots 4-in./10-cm.	Baskets 10-in./25.5-cm.	Pot 4-in./10-cm.	Basket 10-in./25.5-cm.						
11–12	—	14	—	1	—	No	2	6–8 in.* 15–20 cm.	No	F. Sun	H
11–12	—	14	—	1	—	No	2	12–15 in.* 30.5–38 cm.	No	F. Sun	H
13–14	—	16	—	1	—	No	2	6–8 in.* 15–20 cm.	No	F. Sun	H

However, they need winter protection to come back in the spring. Even with protection, though, re-emergence isn't completely assured. Although dusty millers are not sold for their flowers, plants produce yellow blooms on upright stems the second year after seeding.

In the southern U.S.: Allow 8 to 10 weeks for foliage sales in packs and 11 to 12 weeks for sales in 4-in. (10-cm.) pots, using 1 plant per pot. Dusty millers are ideal for landscape and container plantings in the South, and overwinter best in the milder winter regions.

*Heights given refer to foliage rather than flower heights.

Green Packs	Flower Packs	Pots 4-in./10-cm.	Baskets 10-in./25.5-cm.	Pot 4-in./10-cm.	Basket 10-in./25.5-cm.	Pinching	Growth Regulators	Garden Height	Staking	Location	Tender/ Hardy
7–9	8–10	—	—	—	—	No	—	12–18 in. 30.5–46 cm.	No	F. Sun	T

during late winter and spring in the southern U.S. Transplants available from November to January will flower until May. High temperatures will weaken and kill the plants in the most extreme environments. Poppies often reseed as well.

Recommended varieties: The orange-flowering variety is the most common selection of California poppies. However, there are a number of flower colors available as separate colors or in mixtures. There are semi-double to double mixtures available as well, though it is the single-flowering forms that are most popular.

Green Packs	Flower Packs	Pots 4-in./10-cm.	Baskets 10-in./25.5-cm.	Pot 4-in./10-cm.	Basket 10-in./25.5-cm.	Pinching	Growth Regulators	Garden Height	Staking	Location	Tender/ Hardy
—	—	—	—	1–2*	—	No	—	—	No	—	T

Class	Type	Seed for 1,000 plants (oz.)	Number of Seeds	Germination Temperature		Lighting	Days To Germinate	Days Sowing To Transplant	Growing On Temperature	
				Fahrenheit	Celsius				Fahrenheit	Celsius
FLOWERING CABBAGE Brassica oleracea	A, C	1/4	7,000/oz. 245/g.	68°	20°	C	7–14	10–15	55°–58°	13°–14°

Ball Culture: The basic difference between flowering cabbage and flowering kale is in the leaf shape; cabbage has rounded leaves and kale has fringed leaf tips.

Once seedling are large enough, transplant them up to their leaves. When ready to transplant to 2¼-in. (5.6-cm.) pots, plant them deep, covering part of the stems. Northern growers can plant to the garden in the fall or spring, along with other cole crops. Plants can be spaced on 12-in. (30.5-cm.) centers to fill in, or farther apart to leave room between them.

This crop is ideal for fall landscaping themes and tolerates frost well. The colors of the leaves intensify as nights get cooler (45° to 50° F or 7° to 10° C). Once foliage color shows, fertilizing should be discontinued to hasten the color development of the foliage. Although edible, plants

Class	Type	Seed for 1,000 plants (oz.)	Number of Seeds	Fahrenheit	Celsius	Lighting	Days To Germinate	Days Sowing To Transplant	Fahrenheit	Celsius
FLOWERING KALE Brassica oleracea	A	1/4	7,000/oz. 245/g.	68°	20°	C	7–14	10–15	55°–58°	13°–14°

Ball Culture: See comments for flowering cabbage.

Class	Type	Seed for 1,000 plants (oz.)	Number of Seeds	Fahrenheit	Celsius	Lighting	Days To Germinate	Days Sowing To Transplant	Fahrenheit	Celsius
GAZANIA G. splendens	A	1/4	12,000/oz. 420/g.	70°	21°	C	10–12	20–25	55°–60°	13°–15°

Ball Culture: Avoid overwatering since gazanias are extremely sensitive to crown rot.

Though gazanias perform best as border or pot plants, they can also be used for landscaping with good results in warm, dry locations . Planted with other annuals that prefer similar conditions, gazanias can create eye-catching displays. Space plants 10 in. (25.5 cm.) apart to fill in, or 12-in. (30.5-cm.) apart to leave room between them.

In the southern U.S.: Allow 10 to 11 weeks for flowering pack sales and 13 to 14 weeks for flowering 4-in. (10-cm.) pots, using 1 to 2 plants per pot. Gazanias can be planted out from late March to June for flowering during early summer. Space plants 8 in. (20 cm.) apart in full sun to partial

Class	Type	Seed for 1,000 plants (oz.)	Number of Seeds	Fahrenheit	Celsius	Lighting	Days To Germinate	Days Sowing To Transplant	Fahrenheit	Celsius
GERANIUM, STANDARD Pelargonium x hortorum	A	1/4*	6,000/oz. 210/g.	70°–75°	21°–24°	C	7–10	10–15	60°–65°	15°–18°

Ball Culture: We recommend using larger packs (18 cells per flat) for geraniums, making the general crop times for pots and packs identical. For specific crop times, see the geranium section of the Ball Seed Catalog. Sow late January to early February for flowering packs and pots by May 20 to 25.

This is an excellent landscaping crop, and will fill in well if spaced on 10- to 12-in. (25.5- to 30.5-cm.) centers. For greater air circulation, space geraniums 12 in. (30.5 cm.) apart, except for the Bandit, Diamond, Multibloom and Elite series which should always be spaced at 10 in. (25.5 cm.).

Geraniums are hardy plants in certain areas of the deep South and far West, such as coastal California. Plants tend to reach only 15 to 20 in. (38 to 51 cm.) tall in the South, but are grown as shrubs in California where they get as high as 4 ft. (1.2 m.). In these warmer climates, plant in full sun from March to May for summer flowering.

Recommended varieties: All Ball Seed geraniums offer good performance in packs and pots, as well as in gardens and landscapes. To create a sea of bright blooms, choose the Elite or Bandit series, especially Bandit Red or Cherry Elite. The deepest red geraniums, Orbit Red, Lone Ranger and Pinto Red, combine good performance with darker zoning.

Seed geraniums are available in a number of flower colors and habit types. The most free-flowering series is the Multiblooms, which is available in a number of flower colors on plants to 10 in. tall. Always keep the Multiblooms actively growing since stressing them in a plug tray or cell pack can detract from later performance.

The Bandit and Elite series are in the next height category, growing to around 12 in. tall. Both series do well in the garden and have similar growth habits. Or try Neon Rose, an excellent choice for landscapes and 4-in. (10-cm.) pots. They have brilliant rose blooms no other variety can match. In the coral-salmon range, Cameo provides a sharp contrast with Victoria Blue salvias and Blue Puffs ageratums in mass plantings. Among the rose and white bicolors, free-flowering Hollywood Star is our favorite.

		Crop Time (Weeks)		No. Plants							
Green Packs	Flower Packs	Pots 4-in./10-cm.	Baskets 10-in./25.5-cm.	Pot 4-in./10-cm.	Basket 10-in./25.5-cm.	Pinching	Growth Regulators	Garden Height	Staking	Location	Tender/ Hardy
5–6	—	6 in./15 cm. pots 12	—	6 in./15 cm. pots 1	—	No	—	10–14 in. 25.5–35.5 cm.	No	F. Sun	H

are recommended more for garnishes or decorative accents.

In the southern U.S.: Central and deep southern regions should sow late in the summer or in early fall for winter and early spring crops. Since these annuals cannot tolerate

intense heat in the garden, they are planted more often in the fall than in the spring.

In special markets, select Japanese strains of flowering cabbage and kale, 18 in. (46 cm.) tall, can be used as cut flowers with the lower leaves removed.

5–6	—	6 in./15 cm. pots 12	—	6 in./15 cm. pots 1	—	No	—	10–14 in. 25.5–35.5 cm.	No	F. Sun	H
11–12	12–13	14–15	6–7	2	—	No	—	8–10 in. 20–25 cm	No	F. Sun	H

shade. This crop can be used as a perennial in selected areas of coastal California; however, it will not tolerate any cold weather or frost. Plants are tolerant of salt.

Recommended varieties: Gazanias are divided between dwarf and vigorous types. Dwarfs do best in pack and 4-in. (10-cm.) pot production, sold in full bloom. Among the dwarf mixtures, Chansonette Mix provides the earliest flowering in packs. Another good performer, the Daybreak

series has separate colors with the same dark-ringed eyes as some of the colors in Chansonette Mix, while the Ministar series has clear, unringed eyes.

—	13–15	13–15	17–18	1	5	No	4	12–20 in. 30.5–51 cm.	No	F. Sun	T

We offer a number of other geraniums with excellent performance and brilliant color; see the Ball Seed Catalog for further details.

The majority of seed geraniums are genetically defined as diploid plants–having two sets of chromosomes. There are also two varieties of tetraploid types which have four sets of chromosomes. Tetraploid plants tend to have a larger leaf with a distinctly larger flower cluster size, all on plants to no more than 14 in. tall in the garden. The primary difference between diploid and tetraploid types from seed is that

the latter requires 15 to 16 weeks to flower from seed. Tetra Scarlet is a large flowered, irridescent, scarlet-red tetraploid geranium with excellent blooming capacity, while Freckles has a medium pink flower with a prominent rose spot on every petal. Both varieties of tetraploid geraniums do well in mass plantings or in containers.

*Most often sold by seed count rather than by weight.

BEDDING, FLORIST & FOLIAGE PLANTS

Class	Type	Seed for 1,000 plants (oz.)	No. Seeds	Germination Temperature Fahrenheit	Celsius	Lighting	Days To Germinate	Days Sowing To Transplant	Growing On Temperature Fahrenheit	Celsius
GERANIUM, IVY Pelargonium peltatum	A	1/4*	6,000/oz. 210/g.	70°–75°	21°–24°	C	5–10	12–18	60°–65°	15°–18°
GERBERA G. jamesonii	A, P, C	1/4*	7,000/oz. 245/g.	68°–72°	20°–24°	C	10–14	25–30	60°–65°	15°–18°
GLOXINIA Sinningia speciosa	P	1/128= 4,000 seeds	800,000/oz. 28,000/g.	65°–70°	18°–21°	L	14–21	35–42	65°–68°	18°–20°
HELIANTHUS H. annuus (Sunflower)	A	1.5	920/oz. 32/g.	68°–72°	20°–22°	C	5–10	*	62°	16°

Ball Culture: The following culture is based on the Summer Showers variety.

For flowering 10-in. (25.5-cm.) baskets in mid-April, transplant around mid-January from a late-December sowing. Growth retardant may need to be applied once during the growing season to tone up leaves and shorten length between nodes. To produce single-color baskets, grow in 18-cell flats until first blooms appear, then transplant.

Ball Culture: Sow seed in flats, 1/4 to 1/2 in. (.6 to 1 cm.) apart, and cover with finely shredded sphagnum peat moss. For blooming 4-in. (10-cm.) pot plants in early May, sow no later than December 1. This is especially important for series such as Festival, Happipot Mix, Tempo and Pandora, which can take up to 24 weeks to develop. Gerberas can be used as both bedding plants and cut flowers.

For cut flowers in September, sow in March and bench in June. These varieties are not as free-flowering as pot types and don't come into full color until 2 weeks later.

In the southern U.S.: For pot or bedding plant production, sow in early December. Allow 18 to 21 weeks for salable, 4-in. (10-cm.) blooming plants by late March or mid-April.

Ball Culture: *After germination, transplant seedlings into 2 1/2-in. (6-cm.) pots. Move up to 5 or 6-in. (13 or 15-cm.) pots with equal parts of soil, peat and sand, and grow on at 65° to 70° F (18° to 21° C). Allowed 5 to 6 months for flowering pot sales, this daylength-neutral crop can be produced year-round.

We recommend that the first flower be pinched out, along with any leaves that develop in the crown of buds before blooming. This will permit light to uniformly develop buds into a strong first show of color. No other special treatments are required.

GOMPHRENA
G. globosa
See the Cut Flowers section.

Ball Culture: Dwarf selections of sunflowers have been around for a number of years. They are the best varieties on which to use the methods provided here. It's better to sow seed direct to the garden or field when using the taller varieties (3 ft. and above). Plants will develop a stronger root system which is able to support the plant. The chart above and the following cultural information is provided as a guide when growing dwarf sunflowers for use as a pot plant for spring flowering sales, exhibition displays, and in other instances where direct sowing to the garden is not an option.

Plant 2 seeds per 4 1/2-in. or 5-in. pot and thin to 1 plant. Grow in a warm greenhouse where the night temperatures stay above 55° F for best effect. If using for exhibition in displays and other conservatory programs, keep in mind that the plants grow erect without basal branching. Therefore, it is suggested that you display a number of plants together to hide the pot below or insert the flowering sunflower pot into a larger container where a creeping or trailing plant is used to accent and hide the lower part of the plant.

In the field, seed sown direct in mid-May will start to flower by the last part of July and, if sowing a wide selection of varieties, will continue until the end of August. Sunflowers will complete the process from open bud to full bloom in a short period of time—from 5 to 10 days depending on the variety. Sunflowers are not noted for long-term color and as seed begins to ripen, the flowers will all dry up and fall away. If growing for seed, keep a watchful eye out for birds, especially Goldfinches, feasting on the seed.

BEDDING, FLORIST & FOLIAGE PLANTS

		Crop Time (Weeks)		No. Plants							
Green Packs	In FlowerPacks	Pots 4-in./10-cm.	Baskets 10-in./25.5-cm.	Pot 4-in./10-cm.	Basket 10-in./25.5-cm.	Pinching	Growth Regulators	Garden Height	Staking	Location	Tender/ Hardy
—	—	14–15	16–17	1	5–6	No	4	12–15 in. 30.5–38 cm.	No	F. Sun to P. Shade	T

Summer Showers performs best in baskets or containers and reaches a height of 12 to 15 in. (30.5 to 38 cm.).

The Tornado series is the newest addition to Ivy leaf types from seed. It is also the first Ivies from seed available in separate flower colors. The foliage holds closer to the basket or pot than

Summer Showers and is up to 2 to 3 weeks earlier to flower.

*Most often sold by seed count rather than by weight.

—	—	22–24	—	1	—	No	1	10–12 in. 25.5–30.5 cm.	No	F. Sun	T

Outdoors, gerberas should be planted in well-drained areas in the spring or early fall. In some parts of the deep South, plants will continue to flower all year round. Gerberas grown for cut flower production on a bench during the

summer months can heat stall and die. Some work has been done on growing gerberas in the field under saran or shade with favorable results.

Note: Don't plant crowns of gerberas too deep in the bed, pot or bench since plants are suceptible to crown rot and can die as a result.

*Most often sold by seed count rather than by weight.

—	—	*	—	1	—	Yes	6	—	—	—	T

—	—	7–8	—	1	—	No	—	12 in. to 10 ft. 30.5 cm.–3 m.	No	F. Sun	T

In the southern U.S.: Outside of growing potted plants for use as novelties, conservatory displays or unusual pot plants, sunflowers should be sown directly to the ground where they will flower during the summer. Plants sown where they are to flower usually bloom at about the same height as their northern counterparts.

Recommended varieties: Sunflowers are not in bloom for a long period of time—only about 14-21 days. Therefore, selecting the variety which meets your needs is very important. At the time of this writing, one of the best-known dwarf varieties is Teddy Bear, a 38-40-in. plant with golden-yellow double blooms. Flowers measure up to 5 in. across and the resulting seed head will often grow an additional 1-2 in. Another dwarf variety is Big Smile, which grows to between 28-35 in. tall with single flowers that bloom earlier than Teddy Bear. In the tall category there is a wide range

of varieties that are either open-pollinated or F_1 in their breeding history. As cut flowers, the pollenless varieties are of the most interest. Varieties like Sunbright bear no pollen which increases the longevity of the cut flower since the blooms don't die so quickly. Sunbright gets 7 ft. tall with 7 1/2-in. single yellow blooms.

*Sunflowers resent transplanting and are short-lived after going through the process. Sow the seeds directly to pots or to the garden for best effect.

Class	Type	Seed for 1,000 plants (oz.)	Number of Seeds	Germination Temperature Fahrenheit	Celsius	Lighting	Days To Germinate	Days Sowing To Transplant	Growing On Temperature Fahrenheit	Celsius
HELICHRYSUM H. bracteatum *See the Cut Flowers section.*										
HELIOTROPIUM H. arborescens (Heliotrope)	A	1/32	53,000/oz. 1870/g.	70°–72°	21°–22°	C. Lt.	4–8	14–21	62°–65°	16°–18°
HYPOESTES H. phyllostachya	A, F	1/8	18,000/oz. 630/g.	70°–75°	21°–24°	C. Lt.	7–10	14–18	60°–65°	15°–18°
IMPATIENS I. wallerana	A	1/32*	40,000– 60,000/oz. 1,400–2,100/g.	75°–78°	24°–25°	C. Lt.	10–18	15–20	58°–60°	14°–15°
KALANCHOE K. blossfeldiana	P	1/256 = 5,860 seeds	2,500,000/oz. 87,500/g.	70°	21°	L	10–15	35–42	60°	15°

HELIOTROPIUM

Ball Culture: The following culture is based on the standard series Marine. Blue Wonder will flower up to a week earlier.

Sowings made in mid-April will be salable green by the third week in June when grown at 60° F nights in a 32-cell tray. When planted to the garden, the plants will start to flower by the last week of July. It's easy to have blooming plants by June by sowing the seed earlier; just remember to keep the temperatures warm around the developing plants in either the greenhouse or cold frame or they will stall.

Marine does not flower uniformly and it may take up to 2 weeks to get all the plants into bloom from the time the first plants start to flower until the final plant blooms. However, once blooming begins, plants will remain in color as long

HYPOESTES

Ball Culture: Hypoestes are sold for their color-splashed foliage rather than their flowers. The green leaves are splattered with colors of white, pink, burgundy, or red. Plants have insignificant blooms which appear late in the year (late August or September from a June planting to the garden) in primarily blue or white. Grown as pot plants in the greenhouse, this crop performs better when growth retardant is applied to keep the main stem in check and uniform with the rest of the breaks.

Space hypoestes 12 in. (25.5 cm.) apart in the garden. Hypoestes can take full sun or, if taller plants are preferred, lightly shaded areas.

IMPATIENS

Ball Culture: While light aids germination, uniform soil temperature and moisture are just as important. We recommend covering seed lightly with sowing medium unless frequent misting can be done daily, as with an automatic misting system. Once in the garden, impatiens vary in height, depending on the light, fertilizer and water provided.

In the southern U.S.: Allow 8 to 9 weeks for flowering packs and 12 to 13 weeks for blooming 4-in. (10-cm.) pots. Space 10 to 12 in. (25.5 to 30.5 cm.) apart in shaded areas, closer if in partial sun. To our knowledge, no impatiens variety performs particularly well in a full-sun planting. The upright varieties, such as the Blitz and Super Elfin series, tend to do better than other types in somewhat sunny locations.

Recommended varieties: These plants make excellent choices for flowering baskets, containers and shaded landscapes, and are great for pack production. Other than the reds and oranges, most flowers are pastel-colored.

KALANCHOE

Ball Culture: Seed is very fine and difficult to handle. Sow on well-drained media and don't cover. Water seed into media using mist irrigation. Transplant seedlings into cell packs or 2¼-in. (5.6 cm.) pots. When plants begin to crowd, transplant to final 4 to 6-in. (10 or 15-cm.) pots.

Kalanchoes are susceptible to rot, making excess water the greatest problem in growing. During the summer, keep plants in a greenhouse or covered cold frame sprayed with whitewash to reduce light intensity. Whitewashing also reduces the need for frequent watering.

March sowings will produce 4-in. (10-cm.) plants for Christmas sales. Start short days September 1 until buds begin to show for flowering sales by mid-December. For earlier flowering, start shading August 1.

BEDDING, FLORIST & FOLIAGE PLANTS

Green Packs	FlowerPacks	Crop Time (Weeks) Pots 4-in./10-cm.	Baskets 10-in./25.5-cm.	No. Plants Pot 4-in./10-cm.	Basket 10-in./25.5-cm.	Pinching	Growth Regulators	Garden Height	Staking	Location	Tender/ Hardy
8–9	—	—	—	—	—	No	—	10–13 in. 25.5–33 cm.	—	F. Sun	T

as the temperatures are warm. As the flowers age they also fade from a royal blue to a bluish-white color. Dead-heading is suggested but not necessary.

Recommended varieties: Marine has been around for decades and is the variety most commonly grown from seed. Heliotrope is often propagated by stem cuttings and a number of selections in trade can be traced back to Marine. Characterized by royal-blue flowers on plants 12-13 in. tall

in the garden, Marine fills in well and can be used in landscapes or as a pot plant during the summer.

Green Packs	FlowerPacks	Pots 4-in./10-cm.	Baskets 10-in./25.5-cm.	Pot 4-in./10-cm.	Basket 10-in./25.5-cm.	Pinching	Growth Regulators	Garden Height	Staking	Location	Tender/Hardy
10–11	—	11–13	14–15	1–2	5–6	No	8	10–18 in. 25.5–46 cm.	No	F. Sun to P. Shade	T

Recommended varieties: Pink Splash has been the primary variety available. It is basal-branching to 15–18 in. tall with prominant pink splotches across the leaf surfaces. Recently, a white-leaved counterpart, White Splash, was introduced. A dwarf version, Pink Splash Select, was then

introduced and grows to 10 in. or less tall in the garden. However, in the cell pack it is often a third to a quarter the size of standard Pink Splash and takes longer to develop in the cell pack when compared to the standard varieties. Pink Splash Select has the most pink coloring in the leaf

of any variety. The Confetti series is highlighted by four colors, including Rose, White, Burgundy and Red. The foliage color of Burgundy and Red are probably too close in color to distinguish between the two.

Green Packs	FlowerPacks	Pots 4-in./10-cm.	Baskets 10-in./25.5-cm.	Pot 4-in./10-cm.	Basket 10-in./25.5-cm.	Pinching	Growth Regulators	Garden Height	Staking	Location	Tender/Hardy
8–9	10–11	13–14	15–16	1	5–6	**	3, 5, 6***	6–14 in. 15–35.5 cm.	No	P. Shade	T

Impatiens vary greatly in flower size, color, height and habit. The Blitz 2000 series and the Showstoppers were bred for basket production, while Super Elfin is an ideal pack and 4-in. (10-cm.) pot performer. Noted for their large blooms and compact plants, the Accent, Dazzler, and Impulse series, are our most dwarf selections. Like all dwarfs, however, they

require a little more time than standard varieties to finish in 10-in. (25.5-cm.) baskets. All Ball Seed impatiens varieties can be used for any of the purposes mentioned above.

*Most often sold by seed count rather than by weight.

**Plants grown in baskets can be pinched; however, most varieties branch freely.

***Impatiens can be kept short by holding back water and feed.

Green Packs	FlowerPacks	Pots 4-in./10-cm.	Baskets 10-in./25.5-cm.	Pot 4-in./10-cm.	Basket 10-in./25.5-cm.	Pinching	Growth Regulators	Garden Height	Staking	Location	Tender/Hardy
—	—	40–45	—	1	—	No	3	—	No	—	—

Class	Type	Seed for 1,000 plants (oz.)	Number of Seeds	Germination Temperature		Lighting	Days To Germinate	Days Sowing To Transplant	Growing On Temperature	
				Fahrenheit	Celsius				Fahrenheit	Celsius
LAVATERA L. trimestris	A	1/2	4,500/oz. 160/g.	72°	22°	C	4–6	*	58°–60°	14°–15°

Ball Culture: Lavatera is a short-lived annual that performs best in the late spring and early summer months from an early spring sowing. Flowers grow to 3 in. across and come in shades of pink, rose, and white.

*Lavatera performs best when sown direct to large-cell plug trays or small containers such as 4-in. pots. Do not attempt to transplant seedlings since the resulting plant is often weak and dies prematurely.

Seed sown direct to 4-in. pots in April, thinned to 1 or 2 plants per pot, will flower between 8-10 weeks later. However, the plants grow fast and often are too vigorous for effective flowering in 4-in. pots, so you should probably

Class	Type	Seed for 1,000 plants (oz.)	Number of Seeds	Fahrenheit	Celsius	Lighting	Days To Germinate	Days Sowing To Transplant	Fahrenheit	Celsius
LEEA L. coccinea	F	2	850/oz. 30/g.	78°	25°	C	28–49	42–70	65°–70°	18°–21°

Ball Culture: One of the faster-growing foliage crops, Leeas finish in 22 to 24 weeks in 4-in. (10-cm.) pots, 1 to 2 plants per pot. For 6-in. (15-cm.) pots, the crop time is 26 to 28 weeks, again with 1 to 2 plants per pot.

LISIANTHUS
See the Cut Flower section.

Class	Type	Seed for 1,000 plants (oz.)	Number of Seeds	Fahrenheit	Celsius	Lighting	Days To Germinate	Days Sowing To Transplant	Fahrenheit	Celsius
LOBELIA L. erinus	A	1/128= 6,640 seeds	1,000,000/oz. 35,000/g.	70°–80°	21°–26°	L	14–20	20–25	60°	15°

Ball Culture: Trailing series are best suited for basket production, while upright types perform well in packs and containers. See the Ball Seed Catalog for specific variety information. We recommend planting lobelias early in the spring or in the summer in cool climates.

Recommended varieties: One of the oldest varieties of all bedding plants, Crystal Palace, still heads the list of recommended lobelias–a bronze leaf variety up to 5 in. tall with deep blue flowers. Blue Moon, the green-leaved counterpart to Crystal Palace, has the same flower color on the same sized plants.

*The number of plants per pot is defined as a clump. Eight to 10 seedlings (or comparable) make up one clump. Each clump fills out a 4-in. pot, while 5 to 6 clumps fills out a standard 12-in. pot.

Class	Type	Seed for 1,000 plants (oz.)	Number of Seeds	Fahrenheit	Celsius	Lighting	Days To Germinate	Days Sowing To Transplant	Fahrenheit	Celsius
LONAS L. annua	A	1/64	99,500/oz. 3500/g.	68°–72°	20°–22°	L	3–6	14–21	60°	15°

Comments: An easy-to-grow annual with golden-yellow flowers on plants that are excellent in an annual border or in containers mixed with other annuals.

Green Packs	FlowerPacks	Crop Time (Weeks) Pots 4-in./10-cm.	Baskets 10-in./25.5-cm.	No. Plants Pot 4-in./10-cm.	Basket 10-in./25.5-cm.	Pinching	Growth Regulators	Garden Height	Staking	Location	Tender/ Hardy
*	*	8–10	—	1–2	—	No	—	16–20 in. 25.5–51 cm.	—	F. Sun	T
—	—	22-24	—	1–2	—	No	—	—	—	—	T
8–9	11–12	12–13	14–15	1*	5–6*	No	—	3–6 in. 7.5–15 cm.	No	F. Sun** to P. Shade	T
8–9	10–11	11–13	—	1–2	—	No	—	10–11 in. 25.5–27 cm.	No	F. Sun or P. Shade	T

sell green or try dwarfing the plant by either cultural or chemical means.

Recommended varieties: There are two varieties that have been available on the market for years and are still common selections today. Mont Blanc is a white-flowering variety, while Silver Cup has a rose-pink flower. Both varieties grow to 16-20 in. tall.

**In full sun locations, plants often flower stall due to heat stress in August. Conversely, fully-shaded plants tend to lack a free-flowering performance.

Class	Type	Seed for 1,000 plants (oz.)	Number of Seeds	Germination Temperature Fahrenheit	Germination Temperature Celsius	Lighting	Days To Germinate	Days Sowing To Transplant	Growing On Temperature Fahrenheit	Growing On Temperature Celsius
MARIGOLD, AFRICAN Tagetes erecta	A, C	1/8–1/4	9,000/oz. 315/g.	72°–75°	22°–24°	C. Lt.	7	10–15	65°–68°	18°–20°

Ball Culture: Often called American marigolds, the robust, African varieties flower best under short days. For sowings after February 15, start short days upon germination and continue for 14 days, maintaining darkness from 5 p.m. to 8 a.m. Short days can actually be started from the time of sowing by using inverted standard flats to cover germination trays. Although African marigolds take up to 2 additional weeks to flower than their French and triploid counterparts, short day treatment helps produce a smaller overall plant habit as well as earlier and more uniform blooming.

These varieties are ideal for 4- to 6-in. (10- to 15-cm.) pot production, using 1 plant per pot. For green pack sales, we suggest 48 to 32 cells per flat; for flowering pack sales, try 18-cell packs. However, African marigolds do best sold green as opposed to selling in full bloom in packs, since tightly grown varieties are susceptible to Botrytis and Rhizoctonia.

Ball Seed's best landscaping series include the Discovery, Voyager, Inca, Marvel and Perfection. Discovery Yellow is

Class	Type	Seed for 1,000 plants (oz.)	Number of Seeds	Germination Temperature Fahrenheit	Germination Temperature Celsius	Lighting	Days To Germinate	Days Sowing To Transplant	Growing On Temperature Fahrenheit	Growing On Temperature Celsius
MARIGOLD, FRENCH Tagetes patula (Boy, Janie, Bonanza, Little Devil series)	A	1/8–1/4	9,000/oz. 315/g.	72°–75°	22°–24°	C. Lt.	7	10–15	65°–68°	18°–20°
Tagetes patula (Queen types)	A	1/8–1/4	9,000/oz. 315/g.	72°–75°	22°–24°	C. Lt.	7	10–15	65°–68°	18°–20°

Ball Culture: Crop times differ from the double-flowered series (such as Boy and Bonanza), to the anemone types (the Queen and Early Spice series), to the single-flowered varieties (Espana Red Marietta) and the royal crested selections (King Tut and Mandarin).

In general, crop times range from 8 to 11 weeks. French and triploid marigolds are the earliest to flower among the Ball Seed varieties. Some will flower in just 7 weeks from a spring sowing; a few of the older Queens and several royal crested types will flower in 10 to 11 weeks.

French marigolds offer the best return for the investment by providing long-term color and excellent outdoor performance. The most popular of these series are the Heros, Boys, Bonanzas, Janies and Little Devils. The Janie and Little Devil series are our most dwarf varieties, while the Boy series features the smallest bloom. The Hero series is

Class	Type	Seed for 1,000 plants (oz.)	Number of Seeds	Germination Temperature Fahrenheit	Germination Temperature Celsius	Lighting	Days To Germinate	Days Sowing To Transplant	Growing On Temperature Fahrenheit	Growing On Temperature Celsius
MARIGOLDS, SIGNATA Tagetes tenuifolia	A	1/4	9,000/oz. 315/g.	72–75	22°–24°	C. Lt.	7	10–15	65°–68°	18°–20°

Ball Culture: Signata varieties have more finely 'feathered' leaves than other types of marigolds and feature a stronger, citrus scent to the foliage. The pleasing fragrance makes signata marigolds ideal for herb gardens when combined with other scented plants.

Germination is noticeably lower than for the French and African types. Signatas do best grown 1 plant per 3 to 4-in. (7.5 or 10-cm.) pot, though typical varieties such as the Gem series don't bloom uniformly.

In the southern U.S.: Plant to the garden once all danger of frost has passed until May, and again in August for flowering until frost. Marigolds perform best in full sun to partial shade.

Class	Type	Seed for 1,000 plants (oz.)	Number of Seeds	Germination Temperature Fahrenheit	Germination Temperature Celsius	Lighting	Days To Germinate	Days Sowing To Transplant	Growing On Temperature Fahrenheit	Growing On Temperature Celsius
MARIGOLD, TRIPLOID Tagetes erecta x patula	A	1/4	9,000/oz. 315/g.	72°–75°	22°–24°	C. Lt.	7	10–15	65°–68°	18°–20°

Ball Culture: Of all the types of marigolds, the triploid marigolds provide the longest overall color in landscape plantings. Though germination is considerably lower than for French and African types, triploid marigolds offer the advantage of quick finishing. The varieties bloom in just 7 to 8 weeks. Plants are best grown in 3 to 4-in. (7.5 or 10-cm.) pots, 1 plant each. In the garden, space triploids on 12 to 15 in. (30.5 to 38 cm.) centers.

In the southern U.S.: Allow 7 to 8 weeks for flowering packs, 10 to 11 weeks for flowering 4-in. (10-cm.) pots. Plant out after all danger of frost has passed until May, and again in August for flowering until frost. Triploids provide the longest continuous performance in plantings, spaced 12 in. (30.5 cm.) apart in full sun to partial shade.

LEGEND • TYPE: A–Annual   P–Pot Plant   F–Foliage   C–Cut Flower   Pe–Perennial

Crop Time (Weeks)				No. Plants		Pinching	Growth Regulators	Garden Height	Staking	Location	Tender/ Hardy
Green Packs	FlowerPacks	Pots 4-in./10-cm.	Baskets 10-in./25.5-cm.	Pot 4-in./10-cm.	Basket 10-in./25.5-cm.						
9–10	11–12	12–13	—	1	—	No	1, 2, 3, 8	10–30 in. 25.5–76 cm.	*	F. Sun	T

the shortest and Perfection is the tallest; all feature blooms 3 in. (7.5 cm.) across or larger. Planted to the garden or bed, African types perform best on 12 to 15 in. (30.5 to 38 cm.) centers.

In the southern U.S.: Allow 10 to 11 weeks for green pack sales, 11 to 12 weeks for flowering 4-in. (10-cm.) pots. Plant once all danger of frost has passed, until May. Space 12 in. (30.5 cm.) apart in full sun to partial shade. Marigolds planted in spring will flower through November.

*Staking is necessary only for taller, upright types such as Crackerjacks Mixture and the Climax, Gold Coin and Jubilee series. Shorter varieties such as the Perfection, Lady and Galore series occasionally need staking later in the season, but have performed well in Ball Seed trials without support.

6–7	8–9	10–11	10–11	1	5	No	No	8–15 in. 20–38 cm.	No	F. Sun	T
7–8	10–11	12–13	12–14	1	5	No	No	8–15 in. 20–38 cm.	No	F. Sun	T

characterized by 2-in. flowers on plants to 12 in. tall. Dwarfs perform best planted on 8-in. (20 cm.) centers in the garden, while taller varieties require 10 to 12 in. (25.5 to 30.5 cm.) spacing.

In the southern U.S.: Allow 7 to 9 weeks for flowering packs, 10 to 11 weeks for flowering 4-in. (10-cm.) pots. Plant to the garden once all danger of frost has passed until the end of May, and again in August for flowering until frost.

Space 8 in. (20 cm.) apart in full sun to partial shade. Plants may heat stall during the hottest parts of summer.

7–8	9–11	10–11	13–14	1	5	No	No	10–12 in. 25.5–30.5 cm.	No	F. Sun to P. Shade	T

7–8	8–9	10–11	10–11	1	5	No	No	10–12 in. 25.5-41 cm.	No	F. Sun	T

BEDDING, FLORIST & FOLIAGE PLANTS

Class	Type	Seed for 1,000 plants (oz.)	Number of Seeds	Germination Temperature Fahrenheit	Celsius	Lighting	Days To Germinate	Days Sowing To Transplant	Growing On Temperature Fahrenheit	Celsius
MATRICARIA **Chrysanthemum parthenium** *See Perennial Plants section.*										
MELAMPODIUM M. paludosum	A	1/4	5,500/oz. 192/g.	65°	18°	C. Lt.	7–10	14–18	60°–62°	15°–17°
MESEMBRYANTHEMUM M. occulatum	A	1/64	107,000/oz. 3,745/g.	65°–68°	18°–20°	L	7–15	18–20	60°–62°	15°–17°
MIMOSA M. pudica	F	1/2	4,500/oz. 157/g.	80°	26°	C	12–15	—	65°	18°
MIMULUS M. x hybridus	A	1/512 to 1/256	624,000/oz. 21,840/g.	60°–70°	15°–21°	L	5–7	20–25	55°–60°	13°–15°

Ball Culture: The treatment for this crop is the same as for marigolds. Melampodiums are not heavy feeders, so keep fertilizer rates between 100 and 150 ppm of nitrogen. Melampodiums grow quickly and work well as container or bedding plants. Spaced 12 to 15 in. (30.5 to 38 cm.) apart in the garden, plants flower from June until the first frost. Flowering freely, melampodiums produce small, bright blooms. Plants have a vigorous habit, with excellent heat and drought tolerance.

Ball Culture: The culture above is based on mesembryanthemum Lunette; other varieties require additional crop time.

Mesembryanthemums are succulents that tolerate dry conditions, so they shouldn't be planted in areas with high moisture. Plants will rot if watered too heavily. Spaced 10 in. (25.5 cm.) apart, this is an ideal choice for full sun location in dry climates.

Ball Culture: Mimosas' stems are covered with small thorns. Leaves close up at night and when touched; repeated touching damages plants and eventually kills them.

Before sowing, pour hot water over seed. Mimosa branches well, so pinching is not required to promote a full habit.

However, under cool weather and short days (winter production), plants do not branch as readily. Use them as foliage plants, with 2 plants per pot to fill in.

Mimosa are best sold green in pots; plants produce small flowers, but are mainly noted for their foliage.

Ball Culture: The bold flowers hold up well but require long days of 13 hours or more of sunlight to appear. For sowings prior to February 15 in the upper U.S., grow under a mum lighting set-up. Transplant from sowing trays or plug flats direct to final containers.

We recommend this crop for areas with cool, dry summers, as in Pacific coastal regions. In other areas, mimulus plants can be sold in 4 to 6-in. (10 or 15-cm.) pots for spring and early summer color.

*A soft pinch is suggested to encourage branching, especially in 4-in. pots. Cell packs need not be pinched.

| Green Packs | FlowerPacks | Crop Time (Weeks) | | No. Plants | | Pinching | Growth Regulators | Garden Height | Staking | Location | Tender/ Hardy |
		Pots 4-in./10-cm.	Baskets 10-in./25.5-cm.	Pot 4-in./10-cm.	Basket 10-in./25.5-cm.						
6–7	7–8	9–10	—	1	—	No	6	18–20 in. 46–51 cm.	No	F. Sun	T
7–8	10–12	12–14	—	1–2	—	No	No	3–5 in. 7.5–13 cm.	No	F. Sun	T
—	—	24	—	2	—	Yes	No	5–7 in. 13–18 cm.	No	F. Sun	T
7–8	8–9	10–11	11–12	1–2	6–7	Yes*	No	8–12 in. 20–30.5 cm.	No	P. Shade	T

Class	Type	Seed for 1,000 plants (oz.)	Number of Seeds	Germination Temperature Fahrenheit	Celsius	Lighting	Days To Germinate	Days Sowing To Transplant	Growing On Temperature Fahrenheit	Celsius
MIRABILIS M. jalapa (Marvel of Peru, Four-O'Clock)	A	5	325/oz. 11/g.	72°	22°	C	4–6	8–12	60°	15°

Ball Culture: As the common name suggests, Mirabilis flowers in the late afternoon.

The plants grow erect with limited basal branching until planted to the garden. After they have taken root they'll

produce secondary branches which greatly increases bud and flower count.

A sowing made in mid-April and transplanted 10 days later will be salable green in about 8 weeks at 60°–62° F with 1

plant per cell in a 32-cell tray. If planted to the garden by mid-June, plants will flower by the middle of July. Bloom time is variable among plants, though the habit and height are uniform until after flowering. Plants often reseed.

Class	Type	Seed for 1,000 plants (oz.)	Number of Seeds	Germination Temperature Fahrenheit	Celsius	Lighting	Days To Germinate	Days Sowing To Transplant	Growing On Temperature Fahrenheit	Celsius
MORNING GLORY Ipomoea tricolor	A	2 to 3	650/oz. 23/g.	65°–70°	18°–21°	C	5–7	*	60°–62°	15°–17°

Ball Culture: Soak seed overnight before sowing. Morning glories resent transplanting and should be sown direct to final containers—such as Jiffy or 4-in. (10-cm.) peat pots—with 1 to 2 seeds per pot. These are easy-to-grow,

profuse bloomers when planted in low-fertility soil and not overwatered. Morning glories' trailing habit makes them difficult to separate, so they are a poor choice for pack sales. Plants should be grown overhead in the greenhouse

or hanging down from the sidewall. Blooms open in the morning and close in early afternoon.

Class	Type	Seed for 1,000 plants (oz.)	Number of Seeds	Germination Temperature Fahrenheit	Celsius	Lighting	Days To Germinate	Days Sowing To Transplant	Growing On Temperature Fahrenheit	Celsius
NASTURTIUM Tropaeolum majus	A	6 to 7	220/oz. 8/g.	65°–70°	18°–21°	C	10–14	*	65°–68°	18°–20°

Ball Culture: In the garden, avoid applying more than a starter fertilizer to the plants during the growing season. Nasturtiums perform better in a less fertile soil with limited

fertilizer applications. Under higher fertility levels, plants often produce lush foliage at the expense of the flowers.

The flowers that are present tend to be hidden under the canopy of foliage.

Class	Type	Seed for 1,000 plants (oz.)	Number of Seeds	Germination Temperature Fahrenheit	Celsius	Lighting	Days To Germinate	Days Sowing To Transplant	Growing On Temperature Fahrenheit	Celsius
NEMESIA N. strumosa cv. Suttonii	A	1/64–1/32	90,000/oz. 3,150/g.	65°	18°	C	10–14	14–21*	55°–60°	13°–15°

Ball Culture: Fluctuating temperatures will cause erratic germination. Temperatures above 65° F (18° C), even for a few hours, will also inhibit germination. Sow January

through February for blooming 4 to 6-in. (10 or 15-cm.) pots in April and May.

*Since nemesias resent transplanting, sow direct to final containers, grow as a plug or transplant seedlings only once if necessary.

Class	Type	Seed for 1,000 plants (oz.)	Number of Seeds	Germination Temperature Fahrenheit	Celsius	Lighting	Days To Germinate	Days Sowing To Transplant	Growing On Temperature Fahrenheit	Celsius
NEMOPHILA N. menziesii N. maculata	A	1/64–1/32	90,000/oz. 3,150/g.	65°	18°	C	10–14	14–21*	55°–60°	13°–15°

Comments: There are two primary varieties sold in Europe and for use in the western coastal areas of the United States.

One is N. menziesii, available as a mixture as well as a sky-blue-flowering variety, which is sometimes listed as

N. insignis though it is really a selection of N. menziesii. The other variety, commonly called Five Spot, is N. maculata. Five Spot has white flowers with 1 purple dot toward the outer lobe of each petal. In foliage and habit, N. maculata looks similar to N. menziesii. If following the culture methods above, N. maculata will be in bloom in 8 weeks, about

10 days sooner than N. menziesii.

Nemophilas are short-lived annuals best used as pot plants for cool-summer locations. They are native to the west coast and often reseed themselves. They prefer a soil which is moist but not wet.

Crop Time (Weeks)				No. Plants							
Green Packs	FlowerPacks	Pots 4-in./10-cm.	Baskets 10-in./25.5-cm.	Pot 4-in./10-cm.	Basket 10-in./25.5-cm.	Pinching	Growth Regulators	Garden Height	Staking	Location	Tender/ Hardy
7–9	—	—	—	—	—	No	—	18–24 in. 46–61 cm.	No	F. Sun	T

In the southern U.S.: Mirabilis is a heat-tolerant annual or tender perennial in the deep south where it sometimes forms a tuberous root which overwinters. Plants are grown from seed sown direct to the garden or are sold as trans-plants in the spring for color until frost. Plants often reach 30 in. or more in height in the warmest areas of the deep south.

*Roots develop so fast that a 32-cell flat can be outgrown within 5 to 7 weeks if grown under warm temperatures. It's easier to grow this crop using 1 plant per 4-in. pot, though it's sold green since flowering is irregular.

*	*	4–5	7–9	2–3	7–10	No	No	—	**	F. Sun	T

*Morning glories are most commonly sown direct to the garden, but blooming plants can be produced in Jiffy pots with the culture provided.

**Plants need a trellis in the garden.

7–8	8–9	9–10	10–11	1–2	5–6	No	No	10–14 in. 25.5–35.5 cm.	No	F. Sun	T

In the southern U.S.: Allow 7 to 8 weeks for bedding plant sales of flowering peat pots. Plant in full sun to partial shade in late summer for blooming plants until early December, and again in mid-winter for flowering plants through June. Nasturtiums have an upright habit and reach 14 to 18 in. (35.5 to 46 cm.) tall.

*Since nasturtiums resent transplanting, sow direct to final containers, such as peat pots; or sow directly into the garden or flower bed.

—	—	13–15	—	1	—	No	No	10 in. 25.5 cm.	No	P. Shade	T

—	—	13–15	—	1	—	No	No	10 in. 25.5 cm.	No	P. Shade	T

*Nemophilas can be transplanted but would probably per-form just as well if the seed were sown direct to the final container or from a plug.

BEDDING, FLORIST & FOLIAGE PLANTS

Class	Type	Seed for 1,000 plants (oz.)	Number of Seeds	Germination Temperature Fahrenheit	Celsius	Lighting	Days To Germinate	Days Sowing To Transplant	Growing On Temperature Fahrenheit	Celsius
NEPHTHYTIS Syngonium podophyllum	F	5–6	270/oz. 9/g.	75°–80°	24°–26°	C	14–21	30–45	65°	18°

Ball Culture: Sow seed immediately upon arrival; don't store. Seed is highly perishable and germination rate decreases rapidly over the first 3 months. Soak seed for several hours before sowing.

Class	Type	Seed for 1,000 plants (oz.)	Number of Seeds	Germination Temperature Fahrenheit	Celsius	Lighting	Days To Germinate	Days Sowing To Transplant	Growing On Temperature Fahrenheit	Celsius
NICOTIANA N. alata (Flowering tobacco)	A	1/128	250,000/oz. 8,750/g.	70°–75°	21°–24°	L	10–15	20–25	60°–62°	15°–17°

Ball Culture: Keep germination temperatures uniform. Sell green in smaller packs such as 48s and 72s; or sell in larger packs, allowing plants to spread. If plants become rootbound and restricted, they may remain too small to perform well in the garden.

Some older varieties sown before March 1 may require long-day treatment. The culture above is based on the Domino and Nicki series, which have flowered in March from January sowings without additional lighting in our trials.

Among the most distinctive garden annuals available, nicotianas look sharp in 4-in. (10-cm.) pots and in tubs. Plants are self-cleaning (require no dead-heading) which makes them ideal for landscaping. Space nicotianas 12 to 15 in. (30.5 to 38 cm.) apart to fill in.

Class	Type	Seed for 1,000 plants (oz.)	Number of Seeds	Germination Temperature Fahrenheit	Celsius	Lighting	Days To Germinate	Days Sowing To Transplant	Growing On Temperature Fahrenheit	Celsius
NIEREMBERGIA N. caerulea	A	1/128	176,000/oz. 6,160/g.	70°	21°	C. Lt.	6–15	25–34	60°–62°	15°–17°

Ball Culture: Keep soil temperatures above 70° (21° C) during germination. Although niergembergias are tender perennials in warm-winter regions, they are best treated as annuals in severe-winter locations. Plants tolerate cool weather well. Nierembergias live year-round in some areas of California, reaching as high as 18 in. (46 cm.).

Trialed in central Illinois, plants grow no more than 12 in. (30.5 cm.) tall and perform well in borders. In full sun areas, the medium-blue flowers fade slightly; planting in partial shade helps retain color intensity.

The purple flowering Purple Robe, used to be the most common variety in the market. Today the white flowering Mont Blanc has redefined nierembergia as a class. Mont Blanc grows 4 to 7 in. tall, with pure white, 1-in. blooms accented by a yellow eye. Purple Robe has 1-in. flowers, ranging in hue from lavender shading to mid-blue, on

Class	Type	Seed for 1,000 plants (oz.)	Number of Seeds	Germination Temperature Fahrenheit	Celsius	Lighting	Days To Germinate	Days Sowing To Transplant	Growing On Temperature Fahrenheit	Celsius
NOLANA N. paradoxa	A	1/4–1/2	5,000/oz. 175/g.	68°–72°	20°–22°	C. Lt.	5–8	15–20	60°–62°	15°–17°

Ball Culture: Displaying medium-blue, or sometimes white, flowers with prominant white throats, these dwarf plants can spread up to 20 in. (51 cm.) across. If producing hanging baskets, it's best to grow nolanas in cell packs, then transplant flowering plants into baskets to keep them from stretching. No studies have been made on the use of growth regulators on this crop; however, it may be wise to consider using them to keep nolanas in shape and looking good.

Class	Type	Seed for 1,000 plants (oz.)	Number of Seeds	Germination Temperature Fahrenheit	Celsius	Lighting	Days To Germinate	Days Sowing To Transplant	Growing On Temperature Fahrenheit	Celsius
OXYPETALUM O. caeruleum Sometimes called Tweedia, its former botanical name	A	1/4	8,000/oz. 286/g.	70°–72°	21°–22°	L	6–10	18–25	60°	15°

Comments: Oxypetalum is a perennial in the farthest areas of the deep south but is treated as an annual in our northern gardens. This unusual plant has interesting foliage with sulphur blue flowers that are speckled with dark blue. Plants are excellent in locations where they get full sun and warmth.

Sowings made in early March will be salable green with buds by late May and in 30–50% bloom by around the 10th of June. Plants are not overly free-flowering but they send out color all summer long.

Crop Time (Weeks)				No. Plants								
Green Packs	FlowerPacks	Pots 4-in./10-cm.	Baskets 10-in./25.5-cm.	Pot 4-in./10-cm.	Basket 10-in./25.5-cm.	Pinching	Growth Regulators	Garden Height	Staking	Location	Tender/ Hardy
—	—	14–18	20–24	2	5–7	No	1	—	—	—	—
6–8	9–10	10–11	—	1	—	No	6	12–15 in. 30.5–38 cm.	No*	F. Sun	T

In the southern U.S.: Allow 5 to 7 weeks for green packs, 8 to 9 weeks for flowering 4-in. (10-cm.) pots. Plant in the garden from March to April for flowering plants until June. Space 12 in. (30.5 cm.) apart in full sun.

Recommended varieties: Starship, available in five colors, is the most dwarf on the market, growing only 10–12 in. tall. The Domino series has 6 colors available on plants to 14 in. tall, while Nicki is available in a variety of colors as well but grows to 18 in. tall.

*The Nicki series are the only varieties which may require staking, especially late in the season or after periods of excessive rain.

8–9	10–11	12–13	14–16	1–2	5–7	No	No	6–10 in. 15–25.5 cm.	No	F. Sun to P. Shade	T/H

somewhat unkept plants. They grow as tall as 10 in.

There are also some vegetative varieties, with vivid violet flowers, generally found on the West Coast. Plants are hardy in Meditteranean climates but die out in severe winters.

—	9–10	10–11	—	1–2	5–7	No	—	10–14 in. 5.5–35.5 cm.	No	F. Sun	T

Space plants 12 to 15 in. (30.5 to 38 cm.) apart in the garden. Nolanas display heat tolerance similar to marigold and

petunias. Both flowers and foliage can fold up partially at night, though not as tightly as portulacas.

10–12	12–13	14–15	—	1–2	—	No	—	15–17 in. 38–43 cm.	—	F. Sun	T

BEDDING, FLORIST & FOLIAGE PLANTS

Class	Type	Seed for 1,000 plants (oz.)	Number of Seeds	Germination Temperature		Lighting	Days To Germinate	Days Sowing To Transplant	Growing On Temperature	
				Fahrenheit	Celsius				Fahrenheit	Celsius
PALM Chamaedorea elegans	F	12	200/oz. 7/g.	75°	24°	C	—	—	65°	18°

Ball Culture: *Sow immediately; do not store.* Seed is perishable and should be used within a few weeks of shipment. Cover seed with 1/2 in. (1 cm.) of germinating medium, maintaining uniform moisture and 75° F (24° C) temperature for best results. Germination is slow and irregular.

When sowing, leave additional space between seeds so transplanting emerging plants to 2 1/4-in. (5.6-cm.) pots doesn't disturb still-germinating seeds.

Allow 45 to 70 weeks to sell green in 4-in. (10-cm.) pots, 2 plants per pot. In general, the time from transplanting to finishing is usually 24 to 28 weeks.

Class	Type	Seed for 1,000 plants (oz.)	Number of Seeds	Germination Temperature		Lighting	Days To Germinate	Days Sowing To Transplant	Growing On Temperature	
PANSY Viola x wittrockiana	A, Pe	1/16–1/8	20,000/oz. 700/g.	65°–75°	18°–24°	C. Lt.	7–10	15–20	50°–55°	10°–13°

Ball Culture: Varieties differ by genetic makeup, flower size and blotched vs. unblotched faces. Genetically, pansies are available as F$_1$, F$_2$ or open-pollinated strains. F$_1$s are the most heat tolerant and have the longest season of color. The F$_2$s and, especially, the open-pollinated strains get leggy as the heat increases; however, they also offer the most unusual flower colors of the Old Swiss Giant Types. Pansies are tender perennials that may survive mild winters. In regions with severe cold, pansies do best when

covered for the winter, though there is no guarantee plants will come back in the spring. Pansies prefer cool weather and tolerate frost.

Good for pack and container production, this crop makes an ideal border or landscaping item, particularly when combined with cool season crops such as snapdragons, calendulas and dianthus. Space 8 to 10 in. (20 to 25.5 cm.) apart in the garden to fill in, farther apart for greater air circula-

tion. To fill out containers, use 2 plants per 4 1/2-in. (10-cm.) pot or larger.

In the southern U.S.: Allow 13 weeks for flowering packs, 14 weeks for flowering 4-in. (10-cm.) pots. In the garden, plant from late summer to December for blooming plants through June. Space 8 in. (20 cm.) apart for best effect. Though good performers in cool weather, pansies cannot tolerate high, continuous heat.

Class	Type	Seed for 1,000 plants (oz.)	Number of Seeds	Germination Temperature		Lighting	Days To Germinate	Days Sowing To Transplant	Growing On Temperature	
PENTA P. lanceolata	P, A	1/128 = 5,469 seeds	1,000,000/oz. 35,000/g.	70°–72°	21°–22°	L	5–12	21–28	60°–65°	15°–18°

Comments: Pentas from seed used to offer only mixtures of lavender, pink, white and rose. However, a new penta, New Look, offers a medium pink flowering plant, which grows as tall as 10 in.

Although plants are easily produced in either 4- or 6-in. pots, the smaller pots are probably most profitable considering crop time and overall performance.

Vegetative cuttings, shipped as rooted liners, offer a wider range of colors and are the most practical for northern growers. Pentas flower under warm (65° F) conditions and long days.

Class	Type	Seed for 1,000 plants (oz.)	Number of Seeds	Germination Temperature		Lighting	Days To Germinate	Days Sowing To Transplant	Growing On Temperature	
PEPPERS, ORNAMENTAL Capsicum species	A, P	1/4	9,000/oz. 315/g.	72°	22°	C	7–12	15–20	60°	15°

Ball Culture: Sow from June to no later than mid-July to produce fruiting plants by early December. Unlike Christmas cherries, ornamental peppers are self-pollinating and don't need wind or insects to produce fruit. We recommend keeping plants in the greenhouse or under protection during summer months.

Allow 8 weeks for bedding plant sales of green packs in flats. Grown in pots, plants should be pinched at least once to encourage branching. In landscape plantings, space 10 in. (25.5 cm.) apart to fill in, farther apart for greater air circulation.

In the southern U.S.: Sow in January for green or early flowering plant sales in April. Ornamental peppers will produce fruit in late summer and fall.

Crop Time (Weeks)				No. Plants							
Green Packs	FlowerPacks	Pots 4-in./10-cm.	Baskets 10-in./25.5-cm.	Pot 4-in./10-cm.	Basket 10-in./25.5-cm.	Pinching	Growth Regulators	Garden Height	Staking	Location	Tender/ Hardy
—	—	45–70	—	2	—	No	—	—	—	—	T
12–13	14–15	16–17	—	2	—	No	3	6–8 in. 15–20 cm.	No	F. Sun to P. Shade	H

Recommended varieties: Among our most popular F$_1$ series, the Universals and Melodys have 1¼-in. flowers with both blotched and unblotched faces, while the Crystal Bowls produce small blooms with only clear, solid colors. Finally, the Maxims, with a flower size similar to the previous 3 series, is made up of all blotched faces. Our 2- to 3- in. flowered F$_1$ hybrids include the top-selling Crowns (unblotched), Rocs (blotched and unblotched), Accords (blotched and unblotched) and Regals (blotched). The Majestic Giant and Happy Face series have the largest flowers and darkly blotched faces. For medium to large-flowered F$_2$ series, we suggest the Jokers and Rainbows, both blotched. Typically large-flowered, the attractive Swiss Giant types are blotched, open-pollinated varieties.

10–13	—	16–18	—	1	—	Yes*	8	12–15 in. 30.5–38 cm.	No	F. Sun	T

*It does appear that these plants are photoperiodic and, while they don't require pinching to branch, some suggest a pinch above the 3rd or 4th node to make a bigger plant; however, you will need 2–3 more weeks to sell.

8–9	11–14	16–20	—	1	3	Yes	—	10–18 in. 25.5.–46 cm.	No	F. Sun	T

Class	Type	Seed for 1,000 plants (oz.)	Number of Seeds	Germination Temperature		Lighting	Days To Germinate	Days Sowing To Transplant	Growing On Temperature	
				Fahrenheit	Celsius				Fahrenheit	Celsius
PETUNIA **Grandiflora, Multiflora and Floribunda Single** **P. x hybrida**	A	1/256–1/12	265,000–285,000 oz. 9,275–9,975/g.	75°–78°	24°–25°	L	10–12	15–20	55°–60°	13°–15°
Grandiflora, Multiflora and Floribunda Double **P. x hybrida**	A	1/128	245,000/oz. 8,575/g.	75°–78°	24°–25°	L	10–12	15–20	55°–60°	13°–15°

Ball Culture: Petunias are separated into 3 classes which vary in flower size and weather tolerance. Grandifloras are by far the most popular, and offer the largest flower size available in single hybrid varieties. Multifloras display the smallest blooms, but have the highest disease tolerance. Floribundas are an exclusive development by Ball Seed that combined the disease tolerance and free-flowering performance of multifloras with the flower size of grandifloras. The result—our best-selling Madness series.

Grandifloras and floribundas are the earliest to bloom; order multifloras require another 1 to 2 weeks for flowering sales. Double varieties are also available as grandifloras, multifloras and floribundas and take a little longer to flower than singles.

Reduced germination has been noted with low humidity. If humidity cannot be controlled by other means, seed can be covered *very* lightly with vermiculite to raise humidity levels around the seed.

When growing bicolors, control the environment carefully. High light, temperature and fertility produce blooms with a narrow stripe of color and more white; low light, temperature and fertility produce a wider stripe of color with less white.

Petunias are an extremely versatile crop, ideal for packs, containers, baskets, gardens, borders and mass plantings. In landscapes, space petunias 10 to 12 in. (25.5 to 30.5 cm.) apart to fill in.

In the southern U.S.: Plants are hardy in areas with mild or no winters, doing well in full sun. A hard frost will kill petunias. In a study by Tim Raiford of Louisiana State University, floribundas provided the best performance. Sow in January for pack sales in mid March; plants will flower until August. Space 12 to 15 in. (30.5 to 38 cm.) apart in the garden. At the height of the season, petunias reach 10 to 16 in. (25.5 to 41 cm.) tall.

Recommended varieties: Varieties should be chosen for sharp, bold color and full-season performance; the whites, reds, pinks, roses and selected blues and burgundys are best. Yellow and older double varieties tend to display an unkempt habit and lack the outstanding flower performance of the other colors.

Class	Type	Seed for 1,000 plants (oz.)	Number of Seeds	Germination Temperature		Lighting	Days To Germinate	Days Sowing To Transplant	Growing On Temperature	
PHILODENDRON **P. hastatum**	F	1/32	79,000/oz. 2,765/g.	70°–85°	21°–29°	C	15–20	52–76	65°–70°	18°–21°
P. lundii	F	1/16	25,000/oz. 875/g.	70°–85°	21°–29°	C	15–20	28–36	65°–70°	18°–21°
P. pertussum **(Monstera deliciosa)**	F	10	100–200/oz. 3–7/g.	70°–85°	21°–29°	C	15–20	28–36	65°–70°	18°–21°
P. selloum	F	1/8	12,500/oz. 437/g.	70°–85°	21°–29°	C	15–20	32–40	65°–70°	18°–21°

Ball Culture: Cover seed with 1/8 in. (.3 cm.) of germinating medium. Philodendrons are relatively quick to finish compared to other seed foliage crops.

Green Packs	FlowerPacks	Crop Time (Weeks) Pots 4-in./10-cm.	Baskets 10-in./25.5-cm.	No. Plants Pot 4-in./10-cm.	Basket 10-in./25.5-cm.	Pinching	Growth Regulators	Garden Height	Staking	Location	Tender/ Hardy
10–11	12–13	14–15	16–17	1	5–6	No	2, 3	10–15 25.5–38 cm.	No	F. Sun	T
11–12	15–16	16–18	16–19	1	5–6	No	2, 3	10–15 in. 25.5–38 cm.	No	F. Sun	T

For best overall performance in a series, try the Flash, Ultra or SuperMagic series. For outstanding blue and bicolor patterned petunias try the Falcon series. The veined Daddy series, especially Strawberry and Orchid, are excellent choices for use in landscaping. Finally, the Dream series, a limited selection of colors with a marked increase in disease tolerance, is the newest series on the market in the large-flowered, grandiflora category.

In the standard multiflora selections, many of the older varieties have lost out in consumer preference. Old time favorites like the Joy, Resisto and Plum series have been replaced by the floribunda varieties. Technically a multiflora, the term floribunda is used to describe the larger flowering, more compact varieties available in multifloras.

Heralded by the introduction of Summer Madness in 1983 and a number of other colors in 1985, the Madness series continues to lead the floribunda varieties by offering a full range of colors with flowers from 2½ to 3 in. across. Some of the best varieties in the Madness series includes Sheer (light pink with deeper rose veins), Pink, Spring (a pink version of Summer Madness), Plum, and Sugar. Other varieties include the Celebrity series which offers the same flower size on a plant 2–4 in. taller than Madness. The Primetime and Carpet series offer a number of colors on more compact plants with a smaller flower size than Madness.

In double flowering varieties, grandifloras and multifloras are available as well. Grandifloras are larger flowered with a larger overall habit than the multiflora doubles. With the success of the single flowering Madness series, the new series, Double Madness, was developed. This new series offers four colors, with additional ones planned. Other multiflora types include both the Delight and Tart series. These two look similar, are vigorous and have good outdoor performance.

—	—	22–27	—	1	—	No	1	—	—	—	—
—	—	15-20	—	1	—	No	1	—	—	—	—
—	—	15-20	—	1	—	No	1	—	—	—	—
—	—	20-22	—	1	—	No	1	—	—	—	—

BEDDING, FLORIST & FOLIAGE PLANTS

Class	Type	Seed for 1,000 plants (oz.)	Number of Seeds	Germination Temperature		Lighting	Days To Germinate	Days Sowing To Transplant	Growing On Temperature	
				Fahrenheit	Celsius				Fahrenheit	Celsius
PHLOX–Annual P. drummondii	A	1/8	14,000/oz. 490/g.	60°–65°	15°–18°	C	10–15	20–25*	50°–55°	10°–13°
PORTULACA P. grandiflora	A	1/128	280,000/oz. 9,800/g.	75°–80°	26°–29°	L	7–10	35–40	65°	18°
PRIMULA P. acaulis	P, Pe	1/16	28,000–37,000/oz. 980–1,295/g.	60°–65°	15°–18°	L	21–28	40–45	—	—
P. malacoides	P	1/128	280,000/oz. 980/g.	60°–65°	15°–18°	L	21–28	40–45	—	—
P. obonica	P	1/128	190,000/oz. 6,650/g.	68°	20°	L	10–20	40–45	—	—
P. polyantha	P, Pe	1/16	28,000–35,000/oz. 980–1,225/g.	60°–65°	15°–18°	L	21–28	40–45	—	—

Ball Culture: Phlox are native to the U.S. and grow as wildflowers in the southern states.

*We recommend sowing direct to the final containers, since phlox doesn't perform well if transplanted from a seedling tray. Soil temperatures are critical to germination.

In the southern U.S.: Allow 9 to 10 weeks for flowering packs; sowing can begin in February for blooming plants in May. Space 6 to 8 in. (15 to 20 cm.) apart in the garden and let plants fill in. The varieties noted here are not as heat tolerant as the wild forms.

Recommended varieties: Annual flowering phlox is characterized by varieties bearing 1 of 2 flowering types.

Ball Culture: Sow direct to the final container, thinning to a clump of 3–8 seedlings per cell. A soil temperature of 75° to 80° (26° to 29° C) is critical for good germination. Keep fairly dry and fertilize only when needed. In an environment with ample feed and water, portulacas produce foliage with few or no flowers. If plants grow too tall, cut them back, leaving 2 to 4 in. (5 to 10 cm.) of growth. Plants will reflower within 3 to 4 weeks. The crop time for baskets listed above is based on Wildflower Mixture, and is similar to that for blooming flats. In basket production, allow soil to dry out between waterings to keep foliage in check.

Flowers close in cloudy weather and in late afternoon as the sun reaches the horizon. Portulacas' buds may not open until plants receive the light needed to flower, particularly in northern greenhouses during the darker, early days of spring. This makes it essential to choose the right variety.

In the southern U.S.: Allow 8 to 9 weeks for blooming flats. In late winter, plant out after all danger of frost or cool

Ball Culture: Sow **P. acaulis** and **polyantha** varieties from July to September for flowering 4-in. (10-cm.) pot sales from December to early March. Transplant from germinating trays into 2½-in. (6-cm.) pots (cell packs) filled with light, well-drained mix. Once established in the pack, grow on at 50° F (10° C) nights. Move into 4-in. (10-cm.) pots before plants become root-bound, and grow on at 40° to 45° F (4° to 7° C). The acaulis and polyantha types are not heavy feeders.

Cool temperatures are the key to quality; grown too warm, short flowers may form which are hidden by the foliage. It's important to develop a good rosette of leaves before winter; otherwise, plants will not form a uniform bud set.

P. obconica: Use caution in growing this type of primula—some people develop allergic rashes and other reactions from continual handling of these plants. We recommend wearing gloves, long-sleeved shirts and other reasonable precautions while potting, watering and all procedures that may bring skin into contact with foliage. For March sales of 4-in. (10-cm.) pots, sow in September or October. The obconica types should be grown warmer than other primulas; keep night temperatures no lower than 45° to 48° (7° to 9° C). Nights of 55° F (13° C) or higher can produce large leaves and sparse flowers.

P. malacoides: Allow 6 months for flowering 4-in. (10-cm.) pots. Make sowings no earlier than June; continue until September for flowering 4-in. (10-cm.) pots from December through late January for Valentine's Day sales. *Sowings after October 15 can produce blind plants.* This culture is based on 45° F (7° C) night temperatures. The

Crop Time (Weeks)				No. Plants		Pinching	Growth Regulators	Garden Height	Staking	Location	Tender/ Hardy
Green Packs	FlowerPacks	Pots 4-in./10-cm.	Baskets 10-in./25.5-cm.	Pot 4-in./10-cm.	Basket 10-in./25.5-cm.						
—	10–11	13–14	—	—	—	No	2	8–10 in. 20–25.5 cm	No	F. Sun to P. Shade	T
10–11	12–13	13–14	13–14	—	—	No	No	8–15 in. 20–38 cm.	No	F. Sun	T
—	—	—	—	1	—	No	—	—	—	—	H
—	—	—	—	1	—	No	—	—	—	—	H
—	—	—	—	1	—	No	—	—	—	—	H
—	—	—	—	1	—	No	—	—	—	—	H

Rounded flower petals highlight the Beauty and Palona series and Globe Mixture. The star-shaped blooms of both the Petticoat and Twinkles mixes are dainty and smaller in size than their round petalled counterparts. Of the varieties listed above, the Beauty series is the most vigorous.

Growing quickly and producing a bushy habit, the Beauty series fills in on 10- to 12-in. centers when planted in full sun. The remaining varieties do best when planted to the garden spaced 10 in. apart. One of the best uses for this crop is as a container plant. We've grown them warm in the

cold frame or greenhouse 2–3 plants per 6-in. pot with an excellent show of color all summer long.

weather has passed for blooming plants until October. Space 8 to 10 in. (20 to 25.5 cm.) apart in full sun.

Recommended varieties: The large, double blooms of Sundance Mix and similar varieties are usually the quickest to flower and stay open. The Sundial series also blooms early and offers a good range of separate colors, plus a

mix. Double Mixture and Calypso Mixture display a wide range of colors, but their flowers close up if light is limited. In a different species closer to the true purslane type, Wildflower Mixture is our earliest to flower in baskets and the most vigorous garden performer. The single flowers bloom in rose, yellow and white. However, flowers open

and close more erratically than other types, and Wildflower Mix can reseed itself and become a nuisance.

malacoides varieties are susceptible to crown rot and should be transplanted at the same depth from pot to pot. Plants are not heat tolerant; if the crop has not been started by mid-June, it may be better to wait until August or September.

Primulas make nice outdoor plants in areas with mild winters, especially the veris varieties which hold their flowers above the foliage. In regions where snow falls, primulas can be used as early spring plants until warm weather sets in. Plants must be closely spaced, though, to be effective in

masses. In the Midwest, pansies are a better investment for outdoor plantings, while primulas are better used as flowering spring pot plants indoors.

Recommended varieties: Our best acaulis include the Saga, Pageant, Ducat and Festive series; there are earlier bloomers, but these offer the most outstanding overall performance. The Pageants are slightly earlier than the Sagas, but both have excellent habits. The Ducat series offers the widest color selection. Displaying the largest flowers of any Ball Seed acaulis series, the Festives are the latest of the 4

to bloom and are best used for late-season sales. Among the obconica types, Juno and Cantata are the earliest and best overall performing series we've seen yet. In the malacoides class, Pink Ice and Snow Cone continue to lead the Ball Seed line for earliness. Of the polyantha types, we recommend the Pacific Giant series, an excellent variety which holds flowers upright on strong stems.

BEDDING, FLORIST & FOLIAGE PLANTS

Class	Type	Seed for 1,000 plants (oz.)	Number of Seeds	Germination Temperature Fahrenheit	Celsius	Lighting	Days To Germinate	Days Sowing To Transplant	Growing On Temperature Fahrenheit	Celsius
RANUNCULUS R. asiaticus	P, C	1/32	42,000/oz. 1,470/g.	50°–60°	10°–15°	—	14–21	45–50	50°–55°	10°–13°

Ball Culture: This culture is based on the Bloomingdale series.

Sow in September for flowering 4-in. (10-cm.) pots in March. Germination and early growth are slow; high temperatures can further inhibit germination. After germination, keep the soil temperature at 60° F (15° C) during the day and 45° to 50° F (7° to 10° C) at night to tone up seedlings. In warmer temperatures, plants will stretch. Establish plants in 4-in. (10-cm.) pots, 1 to 2 seedlings per pot, before dropping to the final temperatures. Day/night temperatures of 68°/50° F (20°/10° C) are needed to develop flower buds once plants are strongly established in the final pots.

Don't allow ranunculus plants near mum lights, since extended daylight tends to interfere with bud formation and development. If the spring is warm and plants become too tall, apply B-Nine when buds first become visible. Once

Class	Type	Seed for 1,000 plants (oz.)	Number of Seeds	Germination Temperature Fahrenheit	Celsius	Lighting	Days To Germinate	Days Sowing To Transplant	Growing On Temperature Fahrenheit	Celsius
SALPIGLOSSIS S. sinuata	A, C	1/64	125,000/oz. 4,375/g.	70°–72°	21°–22°	*	12–15	18–25	50°–55°	10°–13°

Ball Culture: Don't cover seed with sowing medium; germination percentages increase when the sowing tray is placed in total darkness. We recommend covering the tray with black plastic or some other material after sowing, or placing the tray in an area without light, but with sufficient warmth to induce germination.

In our trials, sowings made in late February have flowered in mid-June in 4 to 5-in. (10 to 13-cm.) pots. For later sales, sow a small crop in February for finishing in 4 or 5-in. (10 or 13-cm.) pots; then sow another, larger crop in March for May pack sales, using the February sowing as a display to sell the packs.

This crop is excellent for spring cut flowers. Sow in January for cut flower production, planting out in March into a ground bed or upright bench in the greenhouse. Space on 8-in. (20 cm.) centers. Seed can be sown in the fall for a winter-flowering crop in the southern U.S.; this hasn't proven economical in the North due to the limited light.

Class	Type	Seed for 1,000 plants (oz.)	Number of Seeds	Germination Temperature Fahrenheit	Celsius	Lighting	Days To Germinate	Days Sowing To Transplant	Growing On Temperature Fahrenheit	Celsius
SALVIA S. coccinea	A	1/8	15,000/oz. 525/g.	70°–75°	21°–24°	L	5–12	18–26	58°	14°
S. splendens	A	1/4	7,500/oz. 262/g.	75°–78°	24°–25°	L	12–15	14–18	60°	15°
S. farinacea	A, C, Pe	1/16	24,000/oz. 840/g.	75°–78°	24°–25°	L	12–15	14–19	60°	15°

Ball Culture: The splendens salvias—such as Fuego, Red Hot Sally and St. John's Fire—are excellent for green or flowering packs and flowering 4-in. (10-cm.) pots. The Victorias and other farinacea series are best suited for green pack or flowering 4-in. (10-cm.) pot sales. This type requires 2 to 3 weeks longer to bloom in pots than the splendens.

S. coccinea is based on the All America Award winner Lady in Red. While this species has been grown for years as a wildflower, Lady in Red is a refined variety bred for uniform pack and garden performance. It is an easy-to-grow annual with scarlet red flowers on plants to 25 in. tall. Plants develop quickly from seed and the bud will be visible in the crown of the plant within 7 to 8 weeks. Plants often grow erect with secondary branches held close to the stem. These will start to develop about 7 to 8 weeks after sowing as the plant sets bud.

In the southern U.S.: Allow 8 to 9 weeks for flowering packs of dwarf splendens varieties; sell taller types green. Space splendens 10 to 12 in. (25.5 to 30.5 cm.) apart in full sun to partial shade in the garden. Planted from March to May, they will flower until November, though blooming is sporadic in July and August. Splendens cannot tolerate frost.

Allow 8 to 9 weeks for geen pack sales of farinaceas. Hardy in the southern states, these salvias are ideal for cut flowers and landscapes, reaching 2½ to 3 in. (76 to 91.5 cm.) in height. Plant to the bed from September on; once established, farinacea varieties will flower until the next fall. To fill in, space 12 to 15 in. (30.5 to 38 cm.) apart in full sun.

Salvia splendens has a number of series that are daylength-sensitive. Michigan State University has trialed many of the popular selections; copies of their report can be ordered through their extension publications office.

Crop Time (Weeks)				No. Plants								
Green Packs	FlowerPacks	Pots 4-in./10-cm.	Baskets 10-in./25.5-cm.	Pot 4-in./10-cm.	Basket 10-in./25.5-cm.	Pinching	Growth Regulators	Garden Height	Staking	Location		Tender/ Hardy
—	—	25	—	1–2	—	No	6	8–10 in. 20–25.5 cm	*	F. Sun		H

buds appear, they will develop in 2 to 4 weeks at a night temperature of 50° F (10° C).

*Plants tend to get top-heavy in the garden and may require some support.

8–10	—	**	—	2–3	—	No	—	24–36 in. 61–91.5 cm.	Yes	F. Sun		T

In the central U.S., home gardeners have shown little interest in salpiglossis plants, probably because they require night temperatures between 55° and 65° F (13° and 18° C). Plants die in the warmer, 75° F (24° C) and higher night temperatures typically experienced in the Midwest during the summer. Planted to the garden in early May, this crop has been known to tolerate temperatures of approximately

40° F (4° C), if properly hardened off. Early May plantings have produced excellent color in the Chicago area until mid-July when located in partial shade with protection from afternoon sun.

*Don't cover seed with sowing medium; place black plastic over the sowing tray and germinate in an area without light.

**Plants tend to grow erect with little branching—this causes them to become top-heavy and fall over. We recommend selling salpi glossis green in the pack; however, we have had success growing in 4-in. (10-cm.) pots using several plants per pot.

7–8	9–10	11–12	—	1	—	**	No	18–25 in. 46–64 cm.	No	F. Sun		T
7–8	9–11*	11–13	—	1	—	No	1, 2, 7, 8	12–24 in. 30.5–61 cm.	No	F. Sun		T
8–9	—	14–16	—	1	—	No	—	20–24 in. 51–61 cm.	No	F. Sun		T

When growing and planting to the garden or landscape, keep the following in mind: In cell packs, salvias respond exceedingly well to warm or hot temperatures by growing erect without basal branching and then flowering. If salvias are planted to the garden from a cell pack in full bloom, especially if no secondary branches have yet to form, the resulting plant will be unproductive and will often die prematurely. While this holds true for a number of salvia species, S. splendens is more sensitive to this condition.

Recommended varieties: For free-flowering performance, look to Red Hot Sally and Fuego, both among our earliest and most compact salvias. Red Hot Sally is a slightly deeper shade of red. Another good choice, Flare displays red

plumes above deep green foliage. Flare is of medium height and displays a vigor and habit similar to Bonfire but on shorter plants. Flare is earlier to flower and has a longer flower spike. The Empire series is made up of six colors ranging from the traditional purple, red and white to more unusual colors including lilac, salmon and rose. For the tallest plants, try Bonfire or America. Among the latest to bloom, these 2 splendens salvias provide a long season of color, right up until frost. The Victoria series as well as Rhea are excellent for both landscaping and dried cut flowers, offering the best of the farinacea characteristics.

*Tall splendens such as Bonfire and America should be sold green due to their height when in flower.

**If the plant blooms in the pack (which it will do readily), suggest to consumers that they pinch back the bloom as they plant it to the garden. This is especially necessary if people are planting this and any of the S. splendens varieties to the garden once warm to hot weather is already present. Plants recently placed in the garden will often stall if they aren't given the water they need to help establish roots on a flowering plant.

Class	Type	Seed for 1,000 plants (oz.)	Number of Seeds	Germination Temperature Fahrenheit	Celsius	Lighting	Days To Germinate	Days Sowing To Transplant	Growing On Temperature Fahrenheit	Celsius
SANVITALIA S. procumbens	A	1/16	28,000/oz. 980/g.	70°	21°	C	7–10	12–15	60°–62°	15°–17°

Ball Culture: Sanvitalias are easily grown annuals that creep or trail. Spaced 12 in. (30.5 cm.) apart in the garden, plants will fill in and flower until frost once established.

In the southern U.S.: Have plants ready for sale once all danger of frost has passed, until early summer. Space 12 in. (30.5 cm.) apart in full sun. Sanvitalias are tender and will not tolerate frost.

Though sanvitalias are an average bedding plant when used alone in a basket or landscape, the varieties available make excellent combination plantings when used with other annuals.

Class	Type	Seed for 1,000 plants (oz.)	Number of Seeds	Germination Temperature Fahrenheit	Celsius	Lighting	Days To Germinate	Days Sowing To Transplant	Growing On Temperature Fahrenheit	Celsius
SCHEFFLERA S. arboricola	F	1/2	4,000/oz. 140/g.	72°–75°	22°–24°	C	14–21	25–40	65°–68°	18°–20°
Brassaia actinophylla formerly known as S. actinophylia	F	1/4	8,000/oz. 280/g.	72°–75°	22°–24°	C	14–21	25–40	65°–68°	18°–20°

Ball Culture: Cover seed with 1/4 in. (.6 cm.) of germinating medium. The arboricola types are more tolerant of low light levels than their actinophylla counterparts.

Class	Type	Seed for 1,000 plants (oz.)	Number of Seeds	Germination Temperature Fahrenheit	Celsius	Lighting	Days To Germinate	Days Sowing To Transplant	Growing On Temperature Fahrenheit	Celsius
SCHIZANTHUS S. x wisetonensis	P	1/32–1/16	45,000/oz. 1,575/g.	60°–70°	15°–21°	L	7–14	21–28	50°	10°

Ball Culture: Leave seed uncovered. Since light inhibits germination, place flats in darkness.

Often referred to as the 'poor man's orchid', schizanthus is excellent as a pot plant and for early spring bedding plant sales in the North. They have a limited life span as indoor house plants. Schizanthus can also be used for cut flower production but they don't ship well. These annuals do best in cool temperatures, preferring full sun to partial shade outdoors. Plants cannot survive hot, humid summers or the harsh winters of the North, so their acceptance and consumer use is rather limited.

Class	Type	Seed for 1,000 plants (oz.)	Number of Seeds	Germination Temperature Fahrenheit	Celsius	Lighting	Days To Germinate	Days Sowing To Transplant	Growing On Temperature Fahrenheit	Celsius
SMITHIANTHA S. zebrina	P	1/256 = 8,203 seeds	3,000,000/oz. 105,000/g.	65°–70°	18°–21°	L	14–21	—	65°–68°	18°–20°

Ball Culture: Sow in May for January sales in 4-in. (10-cm.) pots. Smithianthas form rhizomes and flower under short days. This crop performs better under greenhouse conditions than in gardens or as houseplants. When the flowering season is over, smithianthas will become weak and go dormant. Allow them to rest for 3 to 4 months, then bring plants back into the greenhouse to grow and flower.

		Crop Time (Weeks)			No. Plants							
Green Packs	FlowerPacks	Pots 4-in./10-cm.	Baskets 10-in./25.5-cm.	Pot 4-in./10-cm.	Basket 10-in./25.5-cm.	Pinching	Growth Regulators	Garden Height	Staking	Location	Tender/ Hardy
6–7	8–9	11–13	—	—	—	No	No	10–14 in. 25.5–35.5 cm.	No	F. Sun	T
—	—	22–26	—	1–2	—	No	1	—	—	—	—
—	—	20–23	—	1–2	—	No	1	—	—	—	—
6–7	—	9–10	—	1	—	—	No	8–10 in. 20–25.5 cm.	No	F. Sun to P. Shade	T
—	—	18–22	—	1	—	—	No	—	—	—	—

Class	Type	Seed for 1,000 plants (oz.)	Number of Seeds	Germination Temperature Fahrenheit	Celsius	Lighting	Days To Germinate	Days Sowing To Transplant	Growing On Temperature Fahrenheit	Celsius
SNAPDRAGON–Annual **Antirrhinum majus** *Also see Cut Flowers section.*	A, C	1/128	180,000/oz. 6,300/g.	70°–75°	21°–24°	L	7–14	15–20	45°–50°	7°–10°
STATICE–Annual **Limonium sinuata** *Also see Cut Flowers and Perennials sections.*	A, C	1/4	8,500/oz. 297/g.	70°	21°	L/C	5–12	12–19	50°–55°	10°–13°
STEIRODISCUS **Gamolepis tagetes**	A	1/64	100,000/oz. 3,500 g.	60°–70°	15°–21°	L	5–6	15–28	55°–58°	13°–14°
STOCK **Matthiola incana** *Also see Cut Flowers section.*	A, C, P	1/8	19,000/oz. 665/g.	65°–75°	18°–24°	L/C	7–14	11–18	50°–55°	10°–13°

Ball Culture: Chill seed for several days before sowing to improve germination. When the dwarf varieties show breaks, move to a cold frame or cold greenhouse and grow on at 45° to 50° F (7° to 10° C) to produce well-branched plants. The culture above is based on flowering dwarf varieties in packs or pots. Allowing 10 weeks, sell medium and tall varieties green.

In the southern U.S.: For dwarf varieties, allow 9 to 10 weeks for flowering pack sales, 13 to 15 weeks for flowering 4-in. (10-cm.) pots. For medium and tall varieties, allow 9 weeks for green pack sales, 13 to 15 weeks for 4-in. (10-cm.) pots showing early color. In late October, when nights remain at 68° F (20° C) or below, plant to the garden through March for flowering plants until June. Snaps do best in full sun.

Recommended varieties: For outstanding dwarf strains, try Tahiti Mixture or the Floral Carpet and Floral Showers series, early pack bloomers that work well in borders and 4-in. (10-cm.) pots. The tallest of these dwarfs, Floral Carpets, stand 8 to 10 in. (20 to 25.5 cm.) high while Tahiti Mixture and the Floral Showers reach only 6 to 8 in. (15 to 20 cm.) with a small spread. A medium-sized series, Sonnet, grows to 24 in. (61 cm.) and produces strong,

Ball Culture: Keep the growing temperatures at 50° to 55° F (10° to 13° C) after the plants are established in the pack or bench. When these temperatures are maintained for 4 to 5 weeks, sinuata will flower earlier than when grown

warmer. Sell green as rosettes. Reaching up to 2 ft. (61 cm.) tall, statice will flower in July from a June planting.

They are excellent as either fresh or dried cuts, as well as

background plants in home or cutting gardens. Statice can also be used in landscaping; however, their open habit requires them to be closely spaced for a full effect. In the garden, plants can be spaced 12 in. (30.5 cm.) apart.

Comments: At the time of this writing there is only one steirodiscus sold throughout the U.S., called Gold Rush. Steirodiscus looks similar to Dahlberg Daisy bearing a similar size flower on plants that look and grow like another form of Dyssodia. However, Gold Rush has purer yellow flowers on plants that are less heat tolerant than Dahlberg Daisy.

In the north central U.S. the plants are often planted to the garden by June 1 as blooming packs or pots, which then flower profusely as long as the weather is mild to warm. However, once our weather turns hot and humid, the plants often look tired and succumb to diseases brought on by the stress of the midsummer heat. Plants should be kept away from brick facades and walls and allowed only to grow

where they are to receive morning and early afternoon sun but are shaded during the hottest part of the day.

Ball Culture: This culture is based on the Midget and Cinderella series and other bedding plant varieties of stock.

Late December sowings will produce flowering 4-in. (10-cm.) pots by mid-March. In the germination tray, the double-flowered types will have serrated-edged leaves, while the single types display smooth-edged leaves.

Both the Cinderellas and the Midgets make good bedding or pot plants for spring sales or fall sales. Single-flowered varieties tend to grow taller and may need staking in the garden. Spaced 10 to 12 in. (25.5 to 30.5 cm.) apart, stocks make excellent border plants. These annuals have little commercial importance for landscaping, except in parts of the deep South and Pacific coastal regions where stocks have a longer season than in the Midwest climate.

In the southern U.S.: Once the hot weather subsides in late summer or early fall, plant to the garden in full sun on 10 in. (25.5 cm.) centers. Stocks prefer cool weather and will flower until the temperatures rise again. This crop has not performed well in mid to southern Texas or southern Florida.

Crop Time (Weeks)				No. Plants								
Green Packs	FlowerPacks	Pots 4-in./10-cm.	Baskets 10-in./25.5-cm.	Pot 4-in./10-cm.	Basket 10-in./25.5-cm.	Pinching	Growth Regulators	Garden Height	Staking	Location	Tender/ Hardy
10–11	15–17*	16–18*	—	1–2	—	—	3, 5	6–36 in. 15–91.5 cm.	**	F. Sun	H

sturdy stems—a perfect choice for landscaping. But among the older, cutting types for the garden, the Rocket series still leads. Standing 30 to 36 in. (76 to 91.5 cm.) tall, the Rockets offer a wide range of colors, along with the largest upright flower stalks and blooms of any snaps other than the greenhouse varieties.

*Crop times are based on dwarf varieties.

**Staking is usually required only on taller types such as the Rockets; however, Coronette, Liberty and the Sonnet series may also occasionally need support late in the season.

8–10	—	—	—	1–2	—	No	—	18–24 in. 46–61 cm.	No	F. Sun	T

In the southern U.S.: Plant from fall to December for flowering through June, spaced 15 to 18 in. (38 to 46 cm.) apart in full sun. Annual varieties are used extensively as dried cut flowers.

*	*	*	*	1–2	5–8	No	—	4–6 in. 10–15 cm.	No	F. Sun	H

*January sowings will produce flowering 32-cell packs 12 to 14 weeks later when grown at 55° F nights. If the temperatures are much warmer, the plants will flower earlier but the plants often fall over in the cell pack and look limp. If warmer temperatures are all that is available, grow in hanging baskets but realize that prime performance will be lost. Regardless of warm temperatures, steirodiscus looks

excellent in packs, 4-inch pots or baskets—especially Belden baskets where individual plants can be grown around the outer side of the container to create a fuller effect.

—	9–10	10–11	—	1	—	No	—	8–10 in. 20–25.5 cm.	No	F. Sun	H

The Midget series is a dark to middle green foliage variety that flowers year round, with limited problems, as long as temperatures are hot. The Cinderellas are more robust with grey-green foliage. The Cinderellas are not winter flowering and perform best when sown as days get longer (late December or January) for late winter and spring color. However, they will flower later than Midget when sown at the same time due to the difference between their habits.

Class	Type	Seed for 1,000 plants (oz.)	Number of Seeds	Germination Temperature Fahrenheit	Germination Temperature Celsius	Lighting	Days To Germinate	Days Sowing To Transplant	Growing On Temperature Fahrenheit	Growing On Temperature Celsius
THUNBERGIA T. alata	A	1.5	1,100/oz. 38/g.	70°–75°	21°–24°	C. Lt.	6–12	—	60°–62°	15°–17°
TORENIA T. fournieri	P, A	1/256	375,000/oz. 13,125/g.	70°	21°	L	7–15	28–35	55°–60°	13°–15°
VERBENA V. x hybrida	A	*	10,000/oz. 350/g.	75°–80°	24°–26°	C. Lt.	10–20	20–25	55°–60°	13°–15°
V. bonariensis	A**	***	115,000/oz. 4025/g.	75°	24°	C. Lt.	5–10	21–28	60°	15°
V. canadensis	A	1/8	12,700/oz. 450/g.	75°	24°	C. Lt.	5–10	21–28	60°	15°
V. rigida	A	***	34,000/oz. 1190/g.	75°	24°	C. Lt.	5–10	21–28	60°	15°

Ball Culture: For best results, sow direct to the final containers. However, thunbergias can be transplanted 12 to 18 days after sowing, if preferred. This vining crop is most often sold in hanging baskets and 3-in. (7.5-cm.) or larger pots. We don't recommend pack sales since plants start to trail quickly. A February sowing will produce flowering 10-in. (25.5-cm.) baskets by mid-April.

Spaced 12 in. (30.5 cm.) apart in the garden, 2 plants per site, thunbergias will trail up trellises and screens. Plants can grow as tall as 5 to 6 ft. (1.5 to 1.8 m.) with this type of support. High temperatures may cause thunbergias to stop blooming, so treat accordingly.

Ball Culture: Sow February 1 for blooming 2¼- in. (pack) sales in early to mid-May. Don't allow soil temperature to drop below 70° F (21° C) during germination.

An old favorite, Fournieri compacta is excellent in landscape plantings spaced on 10 to 12 in. (25.5 to 30.5 cm.) centers. Clown Mixture offers less vigor but blooms in a bright range of colors spaced 10 in. (25.5 cm.) apart.

Clown Mixture requires an additional 1 to 2 weeks for pack and pot sales. Torenias do best in afternoon shade; planted in full sun, the leaves darken but the flowers burn in mid-summer.

Ball Culture—V. x hybrida: Germination is the key to success with verbenas. Chill seed for 1 week before sowing. The night before sowing, water in flats and use a Banrot drench at 1/2 tsp. (7.4 ml.) per gal. (3.8 l.) of water. Sow seed without additional watering in. Cover flat with black plastic until germination begins. Because of their genetic makeup, most verbena varieties are poor germinators; following these guidelines can increase germination percentages to 65% or better. Verbena seed is extremely susceptible to rot when moisture is high.

Crop times differ for spreading and upright types, with the spreading types generally flowering 10 to 18 days sooner than their upright counterparts.

Though both types tend to go in and out of bloom during the summer, the spreading varieties seem to be more free-flowering and less prone to cyclical flowering patterns. The Romance and Valentine series have performed well as border plants in Ball Seed trials, though verbenas are generally shunned in landscaping because of their erratic flowering. To fill in, space upright types 10 in. (25.5 cm.) apart and spreading types 10 to 12 in. (25.5 to 30.5 cm.) apart. We recommend sunny locations to yield more free-flowering plants.

In the southern U.S.: Allow 11 to 12 weeks for flowering pack sales of spreading varieties and early-flowering upright types. Allow 14 to 15 weeks for 10-in. (25.5-cm.) hanging baskets. Planted to the bed in late March, verbenas will flower until August. Space upright types 10 in. (25.5 cm.) apart and spreading types 10 to 12 in. (25.5 to 30.5 cm.) apart in full sun.

Ball Culture—V. bonariensis and V. canadensis: Sowings made in early April will be flowering by late June and will retain their color all season long until the weather turns cold.

LEGEND • TYPE: A–Annual   P–Pot Plant   F–Foliage   C–Cut Flower   Pe–Perennial

Crop Time (Weeks)				No. Plants								
Green Packs	FlowerPacks	Pots 4-in./10-cm.	Baskets 10-in./25.5-cm.	Pot 4-in./10-cm.	Basket 10-in./25.5-cm.	Pinching	Growth Regulators	Garden Height	Staking	Location	Tender/ Hardy
—	—	8–9	10–11	1–2	6–7	*	—	—	Yes	F. Sun	T

*Since they tend to branch along the stem, pinching is unnecessary unless faster branch development is desired.

10–11	12–13	13–15	—	1	—	No	—	6–8 in. 15–20 cm.	No	F. Sun to P. Shade	T

In the southern U.S.: Once all danger of frost has passed in late winter or early spring, plant 8 to 10 in. (20 to 25.5 cm.) apart in full sun to partial shade. Torenias will flower until frost strikes, usually in September. This crop dies in cold weather.

—	12–13	14–15	16–18	1–2	5–6	No	2, 5, 8	8–12 in. 20–30.5 cm.	No	F. Sun	T
9–10	—	11-13	—	—	—	Yes	—	30–36 in. 76–91 cm.	No	F. Sun	H
9–10	10–12	12–14	15-17	1	5	No	2	12–15 in. 30.5–38 cm.	No	F. Sun	H
10–12	—	14-15	—	—	—	No	—	12–14 in. 30.5–35.5 cm.	No	F. Sun	T

Ball Culture—V. rigida: Somewhat slower-growing than the other species types of verbena, V. rigida has a more open habit than does either V. bonariensis or V. canadensis, so you will need more plants for a landscape display than the other two selections. April-sown plants will flower by early June and will remain in bloom until frost.

Regarding V. bonariensis and V. rigida specifically: Neither of these (in particular, V. bonariensis) fit into the normal production of bedding plants for cell-pack sales. Instead, they are best sold green, in 4-in. pots or larger. V. rigida will flower within the 4-in. pot, but V. bonariensis should be in a gallon to be effectively used.

Germinaton rates of 30% or less can be expected on V. bonariensis, and 50 to 70% on older seed of V. rigida. Experiments to break dormancy are not conclusive at the time of this writing.

*Germination rates vary dramatically. For Romance, Norvalis and Amour, you need 1/8–1/4 oz. for 1,000 plants. For all others you will need 1/4 oz. of seed.

**A tender perennial treated as an annual here in the Midwest.

***Seed germination erratic and low seedling stands are common.

Class	Type	Seed for 1,000 plants (oz.)	Number of Seeds	Germination Temperature		Lighting	Days To Germinate	Days Sowing To Transplant	Growing On Temperature	
				Fahrenheit	Celsius				Fahrenheit	Celsius
VINCA Catharanthus roseus	A	1/16–1/8	21,000/oz. 735/g.	75°–80°	24°–26°	C. Lt.	7–15	30–35	65°–68°	18°–20°

Ball Culture: Sow seed in flats and cover lightly with sowing medium; keep in darkness until germinated. Maintain 78° to 80° F (25° to 26° C) temperatures for 3 days, then drop to 75° to 78° F (24° to 25° C) for the remainder of germination. Vincas are very sensitive to overwatering and cool temperatures. Grow plants warm at 65° F (18° C) for the first 3 weeks after transplanting, then drop the night temperature to 60° F (15° C).

In the southern U.S.: Allow 11 to 12 weeks for flowering pack sales, 15 to 16 weeks for flowering 4-in. (10-cm.) pots, 1 plant per pot. Vincas will not tolerate cold weather; after all danger of frost has passed plant to the garden for flowering until frost. In areas with mild winters, vincas should be spaced 15 to 18 in. (38 to 46 cm.) apart in full sun. Plants can reach 15 to 20 in. (38 to 51 cm.) tall.

Class	Type	Seed for 1,000 plants (oz.)	Number of Seeds	Germination Temperature		Lighting	Days To Germinate	Days Sowing To Transplant	Growing On Temperature	
				Fahrenheit	Celsius				Fahrenheit	Celsius
ZINNIA Z. angustifolia (formerly Z. linearis)	A	1/8	14,000/oz. 500/g.	70°–72°	21°–22°	C	4–8	25–35	60°	15°
Z. elegans	A	1/2	2,000–6,000/oz. 70–210/g.	70°–72°	21°–22°	C	3–7	10–15	60°	15°

Ball Culture—Z. elegans: Zinnias can be sown direct to the final container.

In the southern U.S.: Allow 4 to 5 weeks for green pack sales, 8 weeks for flowering dwarf varieties in 4-in. (10-cm.) pots. Medium and tall varieties should be sold green in packs once all danger of frost has passed, for flowering plants until October.

Recommended varieties: The nonhybrid varieties generally flower up to 7 days later than the F$_1$ types and may have a more variable plant habit than their uniform cousins. Popular dwarf varieties such as the nonhybrid Pulcino series and the F$_1$ Peter Pan and Dasher series are excellent as bedding, container and landscaping plants. Though not as uniform as the hybrids, Pulcinos have been successfully used in commercial landscaping, flowering right up until frost.

In Ball Seed trials, the medium and tall series have displayed more disease tolerance in both the greenhouse and garden than the dwarf types. For medium uprights, we recommend Lilliput Mixture which grows from 18 to 24 in. (46 to 61 cm.) tall. For the tallest zinnias, try the Ruffles series or State Fair Mixture, with eye-catching 5 to 6-in. (13 to 15-cm.) flowers—the largest available.

Ball Culture—Z. angustifolia: Spring sowings will be starting to flower 8 to 9 weeks later in a 48-cell tray but the plants will not have filled in yet for another 1-2 weeks. Once flowering begins, the plants will remain in bloom as long as the temperature stays warm. There are two flower colors available at the time of this writing. Classic is the orange-flowering type that has been around for years with blooms to 1 1/2 in. across. Classic White, also called Star White, will put out muted white blooms at first and change to pure white as the season progresses. This muted color is expressed under cool temperatures and/or extended cloudy weather.

| Green Packs | FlowerPacks | Crop Time (Weeks) | | No. Plants | | Pinching | Growth Regulators | Garden Height | Staking | Location | Tender/Hardy |
		Pots 4-in./10-cm.	Baskets 10-in./25.5-cm.	Pot 4-in./10-cm.	Basket 10-in./25.5-cm.						
—	14–15	16–17	18–20	1	5–6	No	1, 3	4–12 in. 10–30.5 cm.	No	F. Sun	T
8–9*	9–10*	12–13	—	1–2	—	***	—	12–14 in. 30.5–35.5 cm.	No	F. Sun	T
5–6*	8–9*	9	—	1	—	No	1,2	6–36 in. 15–91.5 cm.	**	F. Sun	T

*The above culture information on green pack sales is based on most zinnia varieties. However, the flowering crop times are for dwarf varieties. Green pack sales are recommended even for dwarfs, though, since many varieties suffer from Botrytis and other diseases if allowed to flower in the pack or grow tightly spaced in small cell packs.

**Staking is required only for taller varieties which grow as high as 2 ft. (61 cm.) or more.

***Pinching is sometimes done to increase the basal branching performance of the plants.

LEGEND • LIGHTING: L–Light   C–Cover   L/C–Light or Cover   C. LT.–Cover Lightly

Class	Number of Seeds	Germination Temperature		Lighting	Days To Germinate	Days Sowing To Transplant	Growing On Temperature	
		Fahrenheit	Celsius				Fahrenheit	Celsius
ACROCLINIUM **Helipterum roseum**	8,500/oz. 300/g.	70°–72°	21°–22°	L/C	6–10	18–26*	60°	15°

Ball Culture: *Higher-quality crops were produced from seed sown direct to the field rather than transplanted.

However, plug-grown and transplanted material will do well. Individual seedling transplant from an open flat was found to cause problems.

Field growing: Seed planted direct to the field in April or May in the northern U.S. will flower in July and August.

Class	Number of Seeds	Germination Temperature		Lighting	Days To Germinate	Days Sowing To Transplant	Growing On Temperature	
AGROSTEMMA **A. githago 'Milas'** **(Corn Cockle)**	42,000/oz. 1,500/g.	65°–70°	18°–21°	C/L	4–8	*	55°	13°

Ball Culture: This entry from Europe has naturalized itself in grain and wasteland areas and has been so noxious that it sometimes showed up in the harvests of grain seed. The seed of Agrostemma has been considered by some to be poisonous and the plants spread and potential seed set should be avoided

around grazing areas. However, as a cut flower it offers some limited filler accents for arrangements with blooms that last for up to 7 days with a treatment. In mild winter areas such as the Pacific Northwest and selected areas of the southern U.S. the plants have overwintered as tender annuals.

The plant prefers cool climates to grow and develop and the best crop is one that is sown early under cool conditions and allowed to flower here in the Midwest by June. Successive sowings can be done but if sown or transplanted to the field in June in the Chicagoland area or similar

Class	Number of Seeds	Germination Temperature		Lighting	Days To Germinate	Days Sowing To Transplant	Growing On Temperature	
AMMI MAJUS **(Lace or Bishop's Flower)***	40,000/oz. 1,400/g.	Day: 70° Night: 68°	Day: 30° Night: 21°	L/C	7–14	18–22**	60°–62°	15°–17°

Ball Culture: The alternating germination temperatures noted above are often required when seed is improperly stored from season to season or when using old seed. Fresh seed seldom requires these techniques prior to germinating. Use germination temperatures of 72° to 75°F to achieve good seedling stands.

Greenhouse growing: Sow direct to beds in February or March for flowering plants in late May or June. Use night temperatures of 55° to 60°F from the time crop is established until flowering.

Field growing: Ammi majus can be direct sown to the field after all danger of frost has passed up until mid-June in the Midwest. Our last field sowing in mid-June sowing will start to flower around early September and be in full bloom by the 15th to 25th of the month. However, later sowings

Class	Number of Seeds	Germination Temperature		Lighting	Days To Germinate	Days Sowing To Transplant	Growing On Temperature	
ANEMONE **A. coronaria**	56,000/oz. 1,960/g.	59°	15°	C	14–21	40–56	42°–48°	6°–9°

Ball Culture: This crop is recommended for greenhouse production only. Anemones grow best in areas with cold or cool winters, including the northeastern U.S. and coastal

California. Use the 38th parallel north (from Washington, D.C. to St. Louis to San Francisco) as the dividing line. The culture above is based on the Mona Lisa and the Cleopatra

series; complete culture information for anemones is too extensive to condense in this guide.

Class	Number of Seeds	Germination Temperature		Lighting	Days To Germinate	Days Sowing To Transplant	Growing On Temperature	
ASTER **Callistephus chinensis** *Also see Bedding, Florist and Foliage Plants and Perennial Plants sections.*	12,000/oz. 420/g.	70°	21°	—	8–10	15–20	60°–62°	15°–17°

Ball Culture: Asters were one of the prime cut flowers in the early 20th century. Unfortunately, their susceptibility to the disease Aster Yellows has resulted in their replacement by chrysanthemums for cut flower use.

Greenhouse growing: Plants can be grown year-round, but the most profitable crop will probably flower between March and early June around the Chicago latitude. Sowings can be made year-round; seedlings need to be lighted 4 hours each night from September 1 to May 1 or plants will flower prematurely. Detailed information is provided on sowing, transplant-

ing, lighting and probable flowering times in the *Ball RedBook*. In general, sowings made in early October and transplanted 2 to 3 weeks later to the bench will flower in March or April. Use 5 to 5½ months a an approximate crop time for crops sown in the fall, and 4 to 4½ months for crops sown in the spring. Don't forget the incandescent lights!

Green Packs	Crop Time (Weeks)			No. Plants		Pinching	Spacing		Staking		Field Height	Seed Required	
	Flower Packs	Pots 4-in./10-cm.	Baskets 10-in./25.5-cm.	Pot 4-in./10-cm.	Basket 10-in./25.5-cm.		Ghse.	Field	Ghse.	Field		Ghse. For 1,000 Plants	Field For 1,000 Sq. Ft./90-Sq. M.
—	—	—	—	—	—	No	—	10x10 in. 25.5–25.5 cm.	—	No	12–14 in. 30–35 cm.	1/4 oz. 7.1 g.	2–3.5 oz. 56.8–99.4 g.

Harvest the flowers as they open and they will continue to open during the drying process.

| — | — | — | — | — | — | No | — | 6x8 in. | — | No | 13–20 in. | 1/32 oz. .9 g. | 5.5 oz. 156 g. |

climate is weather dependent in regards to the plants performance. In general, a February or March sowing results in transplantable plants between 6 to 10 weeks depending on the cell pack size and number of seedlings used per individual cell. These plants are field or garden planted as soon as the ground can be worked and the night air is still cool but frost is past. In our climate, a cool June will continue to produce flowers until July but if our season turns warm or hot early, the plants live but produce limited flower power all season long.

*Sowing should be made direct to the field or cell pack and repeated transplantings should be avoided.

| — | — | — | — | — | — | No | 12x15 in. 30.5x38 cm. | 6x8 in. 15x20 cm. | 1 tier | *** | 2–3 ft. 61–91.5 cm. | 1/16 oz. 1.8 g. | 1–1.5 oz. 28.4–42.6 g. |

have produced weaker plants in September due to the heat, so the sowings should be in May instead of June.

*Botanical names may be a source of confusion for this class. True Queen Anne's Lace is known botanically as Daucus carota var. carota, and is a noxious weed in a number of states. Ammi majus is less offensive, as well as shorter and smaller flowered than true Queen Anne's Lace.

** Ammi majus is best sown direct rather than transplanted.

***If transplanting from plugs or liners, Ammi majus often requires additional support to maintain an upright habit both in the greenhouse and in the field.

| — | — | * | — | 1 | — | No | 10x10 in. 25.5x25.5 cm. | — | — | No | — | 1/32 oz. 886 mg. | — |

Please contact our Customer Service Department or your Ball Seed salesman for additional cultural information on how to grow this crop.

*When growing in 4-in. (10-cm.) pots, Mona Lisa is considered a 5-month crop during the summer, and a 6-month crop during the winter.

| 7–8 | — | 15–16 | — | 1 | — | No | 6x8/8x8 in. 15x20/ 20x20 cm. | 8x8 in. 20x20 cm. | 2 tiers | 2 tiers | 2–3 ft. 61–91.5 cm. | 1/8 oz. 3.6 g. | 3–3.5 oz. 85.2–99.4 g. |

Field growing: In any area plagued by leaf hoppers (there are many), field growing of this crop is not practical unless protection can be provided. For the more adventurous growers, sow from March 15 to April 15 and transplant 3 to 4 weeks later to cell packs or jiffy pots. Plant in the field around mid-May to early June for flowering late July or mid-August (depending on the sowing date). Asters in the field are grown under cloth to decrease the chance of insects transmitting diseases to the crop.

CUT FLOWERS

Class	Number of Seeds	Germination Temperature		Lighting	Days To Germinate	Days Sowing To Transplant	Growing On Temperature	
		Fahrenheit	Celsius				Fahrenheit	Celsius
BELLS OF IRELAND Moluccella laevis	4,200/oz. 147/g.	Day: 86° Night: 50°	Day: 30° Night: 10°	L	12–21	21–28	60°–62°	15°–17°
CACALIA Emilia javanica cv. Lutea (Tassel Flower)	40,000/oz. 1,400/g.	68°–72°	20°–22°	C	8–15	*	64°–68°	18°–20°
CARTHAMUS C. tinctorius (Safflower)	800/oz. 28/g.	68°–72°	20°–22°	L	5–14	15–21*	65°–68°	18°–20°
CELOSIA C. plumosus	39,000/oz. 1,365/g.	75°	24°	C	8–10	10–15	65°–68°	18°–20°
C. cristata Also see Bedding, Florist and Foliage Plants and Perennial Plants sections.	34,000/oz. 1,190/g.	75°	24°	C	8–10	10–15	65°–68°	18°–20°

Ball Culture: Germination is the most difficult part of growing this crop. If you're having problems, treat the crop as a frost germinator instead. If growing in an area with severe winters, you can sow the seed, cover it lightly, and place the seed flat outside, exposed to winter conditions for 6 to 8 weeks. Then bring it back into the greenhouse and germinate at 70° F (21° C). This isn't a practical method for most growers, but it can be used if chilling the seed in the refrigerator at the recommended temperatures proves inadequate.

Greenhouse growing: Although not recommended, greenhouse growing is possible. Sowings in February or early March will produce flowering plants in May if grown at night temperatures of 50° to 55° F (10° to 13° C). This culture applies to seedlings spaced 4x4 in. (10x10 cm.) and grown single-stem.

Ball Culture: Cacalia is an easy-to-grow annual which produces clusters of small, fiery scarlet or orange flowers on 24-in. (61-cm.) stems. When sown in early spring, plants flower very freely from July until frost. This plant will do very well in almost any soil type in full sun and will often reseed itself. Cacalia should be used fresh.

*Cacalia responds poorly to transplanting, so sow direct to the final bed or field.

Ball Culture: Produced as either dried or fresh everlasting cut flowers, this crop does best when sown direct to the field or greenhouse bed. If to be transplanted, start seedlings in plug trays and don't attempt single seedling separation from an open germination tray.

Two types of safflower are available. Developed in the Orient, the first type has round leaves and blooms earlier than the second type, which is a European strain with an acute leaf shape. The Oriental strain also has fewer thorns along its stems than the European strain and is easier to work with in close quarters.

Greenhouse growing: Sowings in late January or February will flower 13 to 15 weeks later depending on temperatures. Do not grow too cool; 55° to 60° F is acceptable but 60° to 65° F will produce flowering plants sooner.

Ball Culture: Crested types can be grown as cut flowers, through the plume, or feather, types are more commonly used. Sales are primarily for dried cut flowers.

Field Growing: Most often grown in the field, celosias should be sown in plug trays or cell-packs 6 to 8 weeks before setting out. Seedlings transplanted to the field in late May will flower in July and August.

For dried cut flower sales, remove the foliage at harvest and hang plants upside down in a dry and airy, dark location until dry—about 10 days. Shake heads vigorously to loosen the seeds inside. The crop is then ready for sale.

Recommended varieties: For a short-stemmed feather variety, try the 20-in. (51-cm.) Century series. Forest Fire, a 30-in. (76-cm.) variety with deep crimson plumes and crimson foliage, also does well as a cut flower. The Feather series is one of the taller varieties 3 ft. (.9 m.) and over, but flowers later than the Century's or Forest Fire. Among the crested types, the 3- to 3 1/2-ft. tall Chiefs offer a range of separate colors with heads up to 5 in. (13 cm.) across, ideal for filler work. Kardinal, a 10- to 12-in. (25.5- to 30.5-cm.) variety, and Toreador, an 18- to 20-in. (46- to 51-cm.)

Green Packs	Crop Time (Weeks)			No. Plants		Pinching	Spacing		Staking		Field Height	Seed Required	
	Flower Packs	Pots 4-in./10-cm.	Baskets 10-in./25.5-cm.	Pot 4-in./10-cm.	Basket 10-in./25.5-cm.		Ghse.	Field	Ghse.	Field		Ghse. For 1,000 Plants	Field For 1,000 Sq. Ft./90-Sq. M.
8–9	10–11	12–14	—	1–2	—	No	4x4 in. 10x10 cm.	8x8 in. 20x20 cm.	No	No	15–20 in. 38–51 cm.	1/2 oz. 14.2 g.	3–3.5 oz. 85.2–99.4 g.
5–7	—	—	—	—	—	No	8x8 in. 20x20 cm.	8x8 in. 20x20 cm.	Yes	**	20–24 in. 51–61 cm.	1/32 oz. 886 mg.	3–3.25 oz. 85.2–92.3 g.
—	—	—	—	—	—	No	6x6 in. 15x15 cm.	6x8 in. 15x20 cm.	Yes	No	24–36 in. 61–91.5 cm.	2.5–3 oz. 71–85 g.	5.5–7 oz. 156–198.5 g.
7–9	11–12	13	—	1	—	No	6x8 in. 15x20 cm.	10x10 in. 25.5x25.5 cm.	Yes	No	32 in. 81 cm.	1/32–1/16 oz. .9–1.8 g.	1/8 oz. 3.6 g.
7–9	11–12	13	—	1	—	No	6x8 in. 15x20 cm.	10x10 in. 25.5x25.5 cm.	Yes	No	32 in. 81 cm.	1/32–1/16 oz. .9–1.8 g.	1/8 oz. 3.6 g.

Field growing: Field growing is the recommended method for this crop. April sowings direct to the field produce 1 1/2-ft. (46-cm.) flowering plants in June. Beware of late sowings– high soil temperatures can cause irregular germination.

Bells of Ireland can be used in either fresh or dried arrangements. To use as a cut flower, the leaves from between the bell-shaped bracts must be removed upon cutting the stem.

**Plants tend to have weak stems, so use 1 layer of netting in the greenhouse. In the field, harvest blooms as color begins to show, but before stems bend.

Not recommended for greenhouse growing.

Field growing: Sowings made direct to the field in early May will flower during the second or third week of July.

Successive sowings are suggested since plants produce one major burst of color and then die.

Beware of very fertile soil types or high fertilizer applications. Carthamus isn't a heavy feeder and it has been suggested that high fertility levels causes yellowing of foliage (though a number of things can cause this symptom), poor flower development and stretching, especially when grown in the greenhouse under cloudy days for extended periods.

* Crop is better sown direct, or transplanted from plugs.

celosia, display intense red, 10- to 12-in. (25.5 to 30.5-cm.) flower heads. Both are grown single-stem without breaks.

CUT FLOWERS

Class	Number of Seeds	Germination Temperature		Lighting	Days To Germinate	Days Sowing To Transplant	Growing On Temperature	
		Fahrenheit	Celsius				Fahrenheit	Celsius
CENTAUREA **C. cyanus** **(Bachelor's Buttons or Cornflower)**	7,000/oz. 245/g.	65°–70°	18°–21°	C.Lt.	7–14	20–25	50°–55°	10°–13°
C. sauveolens	5,500/oz. 192/g.	65°–70°	18°–21°	C.Lt.	7–14	18–24	50°–55°	10°–13°

Ball Culture: Greenhouse growing: Though not as common as field growing, a greenhouse crop sown in mid-September will flower from March to May if given supplemental lighting from mid-January to mid-March. Use an incandescent lighting system as for mums from 11 p.m. to 2 a.m. each night. For spring flowering in the greenhouse, sow December to mid-January for blooming plants in May.

Field growing: In Florida and the central coastal and southern regions of California, sow seed directly to the field in September for flowering plants from February to June. In the eastern and midwestern states, sow seed in the field April to mid-May for flowering plants from July to September.

Class	Number of Seeds	Germination Temperature		Lighting	Days To Germinate	Days Sowing To Transplant	Growing On Temperature	
CIRSIUM **C. japonicum** **(Japanese Thistle)**	12,250/oz. 429/g.	60°–65°	15°–18°	L/C	7–14	21–28	55°–58°	13°–14°

Ball Culture: Cirsium tolerates poor soil, but often becomes 'weedy' in warm-winter areas. The following schedules are based on the Early Beauty series, which, unlike many cirsium varieties, is daylength neutral.

Greenhouse growing: Plants will flower in November from a June or July sowing. With greenhouse crops, 3 pickings or harvests can be expected approximately November, February and May, although we get only 1 to 2 good pickings here in the Midwest.

Class	Number of Seeds	Germination Temperature		Lighting	Days To Germinate	Days Sowing To Transplant	Growing On Temperature	
CRASPEDIA **C. uniflora**	10,000/oz. 350/g.	72°	22°	L/C	7–14	14–21	65°	18°

Ball Culture: Treat craspedia as a tender perennial in warm-winter areas, and as an annual in the northern and eastern U.S. Craspedia is not frost-tolerant. Most studies have been done on field growing, so little is known about growing craspedia in the greenhouse.

Field growing: Sow seed in the greenhouse in early spring in the Midwest, or November-December in warm-winter areas, for field planting 8 to 10 weeks later. Plants will

Class	Number of Seeds	Germination Temperature		Lighting	Days To Germinate	Days Sowing To Transplant	Growing On Temperature	
EUSTOMA **E. grandiflorum** *See Lisianthus.*								

Class	Number of Seeds	Germination Temperature		Lighting	Days To Germinate	Days Sowing To Transplant	Growing On Temperature	
FREESIA **F. x hybrida**	2,400/oz. 84/g.	65°	18°	C	21–25	*	50°	10°

Ball Culture: Soak seed overnight before sowing to increase germination.

Greenhouse growing: Freesia is suggested for greenhouse production only in the midwestern states, though it can be grown outside under saran in California. Sowings done in the midwest from May to July, benched 5 to 7 weeks after sowing, will flower from January until March under night temperatures of 50° to 55° F (10° to 13° C). If growing only

CUT FLOWERS *(vertical side tab)*

| Green Packs | Crop Time (Weeks) | | | No. Plants | | Pinching | Spacing | | Staking | | Field Height | Seed Required | |
	Flower Packs	Pots 4-in./10-cm.	Baskets 10-in./25.5-cm.	Pot 4-in./10-cm.	Basket 10-in./25.5-cm.		Ghse.	Field	Ghse.	Field		Ghse. For 1,000 Plants	Field For 1,000 Sq. Ft./90-Sq. M.
7–8	9–10	10–13	—	2–3	—	No	6x6 in. 15x15 cm.	4x6 in. 10x15 cm.	Yes	*	2–3 ft. 61–91.5 cm.	1/2oz.** 14.2 g.**	3–3.5 oz. 85.2–99.4 g.
7–8	9–10	10–13	—	2–3	—	No	6x6 in. 15x15 cm.	4x6 in. 10x15 cm.	Yes	*	20–28 in. 51–71 cm.	1/2 oz.** 14.2 g.**	3–3.5 oz. 85.2–99.4 g.
6–7	—	8–9*	—	1	—	No	15x15 in. 38x38 cm.	10x10 in. 25.5x25.5 cm.	1 tier	No	18–24 in. 46–61 cm.	1/8–1/4 oz. 3.6–7.2 g.	3–3.5 oz. 85.2–99.4 g.
8–10	—	—	—	—	—	No	—	15x15 in. 38x38 cm.	—	1 tier	2–3 ft. 61–91.5 cm.	1/4 oz.* 7.1 g.*	3–3.25 oz. 85.2–92.3 g.*
—	—	—	—	—	—	No	4x6 in. 10x15 cm	—	1–2 tiers	—		** 1 oz. 28.4 g.	—

Maintain a minimum night temperature of 55° F (13° C). In field production, plants from seed sown during June and July will produce a few flowers on short plants and then die. Plants started under short days and cool weather (i.e., fall, winter and early spring) will promote basal branching that elongates when given long days (i.e., mid-spring and summer).

*Staking is determined by the wind factor.

**Though centaurea can be transplanted to the field or greenhouse, direct sowing yields higher quality crops and reduces labor expense.

Field growing: In the Midwest, sow seed in late March or April for field transplanting in June. Plants will be in full bloom from early to mid-August.

*4-in. (10-cm.) pots will not be in flower.

flower sporadically by August. It's best to sow early and transplant to the field from either large cell packs or small pots to maximize flower production.

*Cleaned seed.

a limited number of plants for your own use, use 8 to 10 plants per 10 to 13-in. (25.5 to 33-cm.) tub and provide stem support. Pots can be easily moved as needed.

*Transplant direct from sowing flat into the final bench or container.

**Plants get 1 to 2 ft. (30.5 to 61 cm.) tall in the greenhouse.

CUT FLOWERS

71

Class	Number of Seeds	Germination Temperature		Lighting	Days To Germinate	Days Sowing To Transplant	Growing On Temperature	
		Fahrenheit	Celsius				Fahrenheit	Celsius
GODETIA G. whitneyi	37,000/oz. 1,295/g.	70°	21°	L/C	10	11–20	55°	13°

Ball Culture: Greenhouse growing: Godetias do best in cool climates such as the Pacific Coast. In the Midwest or eastern U.S., they can be grown in shallow flats for local greenhouse cut flower spring sales. Godetia can be grown in the summer in the Midwest, but stems are often weak and the plants die from the hot weather. However, in California, godetias do best in the field in full sun. Outside California, grow in greenhouses as early spring crops under cool conditions.

Class	Number of Seeds	Germination Temperature		Lighting	Days To Germinate	Days Sowing To Transplant	Growing On Temperature	
GOMPHRENA G. globosa	**	72°	22°	L/C	10–14	20–25	68°	20°

Ball Culture: Gomphrena is a heat and drought-tolerant plant that can be sown directly to the field or transplanted to the field from cell packs; plants fill in quickly. Flowers are used dried.

Field growing: Gomphrena is recommended only for field growing. In the Midwest, sow directly to the field in April or early May for flowering plants in July.

*Crop times are based on Buddy and other dwarf flowering types. For pack sales of larger varieties, allow 7 to 8 weeks and sell green only. They become too large in the pack or

Class	Number of Seeds	Germination Temperature		Lighting	Days To Germinate	Days Sowing To Transplant	Growing On Temperature	
GYPSOPHILA–Annual G. elegans *Also see Perennial Plants section.*	24,000/oz. 840/g.	70°–80°	21°–26°	L/C	10–15	21–28	50°–55°	10°–13°

Ball Culture: Greenhouse growing: If grown at 50° to 52° F (10° to 11° C) nights, varieties like Convent Garden will flower in about 3 months from sowing. When grown during December and January, crop time will increase to 3½ months for the Midwest. Annual gypsophilas can be grown in 3-in. (7.5-cm.) deep flats by sowing directly into the flat or transplanting into it from a seed flat. Use spacing of 3 x 3 in. (7.5 x 7.5 cm.) or 3 x 4 in. (7.5 x 10 cm.).

Class	Number of Seeds	Germination Temperature		Lighting	Days To Germinate	Days Sowing To Transplant	Growing On Temperature	
HELICHRYSUM H. bracteatum	45,000/oz. 1,575/g.	70°–75°	21°–24°	L/C	7–10	20–26	60°–65°	15°–18°

Ball Culture: Helichrysums grow very quickly as bedding plants. They may be too large in packs by blooming time, so they should be sold green. The bedding plant notes in the chart above are based on dwarf varieties, such as Bright Bikinis Mixture.

Field growing: Helichrysums are recommended only for field growing. In the Midwest, sow seed in the greenhouse in April or May for transplanting to the field 6 to 8 weeks later. Seed can also be sown direct to the field once the danger of frost is past.

Plants will flower from late July to frost. This culture is for the big, tetraploid types up to 3 ft. (91.5 cm.) tall. Harvest flowers when 2 to 4 rows of petals are folded outward and before the center is exposed.

Class	Number of Seeds	Germination Temperature		Lighting	Days To Germinate	Days Sowing To Transplant	Growing On Temperature	
IBERIS I. amara **(Annual Candytuft, Rocket Candytuft or Globe Candytuft)**	10,000/oz. 350/g.	68°–72°	20°–22°	L/C	7–14	16–24	50°–55°	10°–13°
I. umbellata *Also see Perennial Plants section.*	10,000/oz. 350/g.	68°–72°	20°–22°	L/C	7–14	16–24	50°–55°	10°–13°

Ball Culture: There are 2 distinct and valuable cut flower types of iberis. The umbellata class produces large heads of umbel-type flowers in rose, purple, lilac and carmine. The hyacinth or coronaria class (I. amara) has long, white, hyacinth-like flowers on erect stems. To harvest, cut I. umbellata types as they mature. Flowers shatter rather easily and must be cut when ready. I. amara, though, can be left to lengthen over 1 to 2 weeks after they are ready. Of the 2 types, I. amara is the more profitable.

Greenhouse growing: Although not recommended, Midwest growers can make sowings in the fall for flowering in mid-winter. The total crop time is 4 to 5 months. Late January and February sowings produce higher quality crops, however. These plants will flower in late April or

| Green Packs | Crop Time (Weeks) | | | No. Plants | | Pinching | Spacing | | Staking | | Field Height | Seed Required | |
	Flower Packs	Pots 4-in./10-cm.	Baskets 10-in./25.5-cm.	Pot 4-in./10-cm.	Basket 10-in./25.5-cm.		Ghse.	Field	Ghse.	Field		Ghse. For 1,000 Plants	Field For 1,000 Sq. Ft./ 90-Sq. M.
8–9	11	14–16	—	1–2	—	*	6x8 in. 15x20 cm.	12x15 in. 30.5x38 cm.	1 tier	No	12–18 in. 30.5–46 cm.	1/32 oz. 0.9 g.	1.75–2.5 oz. 49.7–71 g.
7–8*	9–10*	13–14	—	1	—	No	—	10x10 in. 25.5x25.5 cm.	No	No	18–24 in. 46–61 cm.	—	**
8–10	—	—	—	—	—	No	3x4 in. 7.5x10 cm.	4x6 in. 10x15 cm.	1 tier	1 tier	18–24 in. 46–61 cm.	1/16–1/8 1.8–3.6 g.	2–3 oz. 56.8–85 g.
6–8	—	10–12	—	1	—	No	—	*	—	1 tier	2.5–3 ft. 76–91.5 cm.	1/32 0.9 g.	2.5–3 oz. 71–85.2 g.
12	—	—	—	—	—	No	6x8 in. 15x20 cm.	8x8 in. 20x20 cm.	1 tier	No	20–24 in. 51–61 cm.	1/8–1/4 3.6–7.1 g.	3–3.5 oz. 85.2–99.4 g.
12	—	—	—	—	—	No	6x8 in. 15x20 cm.	8x8 in. 20x20 cm.	1 tier	No	15–20 in. 38–51 cm.	1/8–1/4 3.6–7.1 g.	3–3.5 oz. 85.2–99.4 g.

Field growing: In mild winter climates, sow seed in the fall, and transplant to the field by October or November. It may be best to sow in the greenhouse and transplant to the field once the plants are strongly established in the containers.

Grow under cool conditions, up to 50° F (10° C). In the Midwest or eastern U.S., field growing of godetias isn't recommended.

*Godetias will branch without pinching. However, pinching upon planting in their final location encourages even blooming rather than 1 branch at a time.

pot to flower. Gomphrenas are not responsive to growth retardants B-Nine, A-Rest or Cycocel.

**The number of seeds per ounce varies from 5,000 for uncleaned seed to 11,500 (402 g.) for cleaned seed. For sowing to the field, 1½ to 2 oz. (42.6 to 56.8 g.) of

cleaned seed are needed, compared to 3 to 3½ oz. (85.2 to 99.4 g.) of uncleaned seed.

Field growing: Seed planted to the field once the ground warms up will flower 2½ to 3 months later. Plants will put out 1 or 2 flushes, so successive sowings made every 3 to

4 weeks will provide continuous color during summer months.

Growers in the Midwest and Northeast have produced crops with better color and habit in the summer when growing under saran, away from the sun.

*Space plants on 10-in. (25.5-cm.) centers within the row, with 2 to 3 ft. (61 to 91.5 cm.) between the rows.

May. Bench the seedlings 2 to 3 weeks after sowing, with 6–8 in. (15–20 cm.) spacing for I. amara types and 8–10 in. (20–25.5 cm.) spacing for I. umbellata varieties. The plants may need 1 or 2 tiers of support. Do not pinch; plants branch readily.

Field growing: For either type, sow seed direct, or start in the greenhouse 6 to 7 weeks earlier. Begin as soon as the field can be worked in the spring, or earlier in mild-winter areas. Spacings for the field will be the same or slightly larger than those noted for greenhouse growing. Field-

grown crops will not be as long-stemmed as their greenhouse-grown counterparts.

CUT FLOWERS

Class	Number of Seeds	Germination Temperature		Lighting	Days To Germinate	Days Sowing To Transplant	Growing On Temperature	
		Fahrenheit	Celsius				Fahrenheit	Celsius
LARKSPUR Consolida ambigua	8,200/oz. 287/g.	55°–65°	13°–18°	C	10–20	28–35	50°–55°	10°–13°
LISIANTHUS Eustoma grandiflorum (formerly E. russellianum)	624,000/oz. 21,840/g.	75°–80°	24°–26°	L	10–20	36–48	See below	15°–18°

Ball Culture: Chill seed for 7 days at 35° F (2° C) for better germination. Like its cousin the delphinium, larkspur is erratic and won't germinate effectively at temperatures above 70°F (21°C). For spring bedding plants, sow seed direct in December or January to gallon containers for sales in May or June. Plants will be close to flowering if not in bloom. Grow on at 55° F (13° C).

Greenhouse growing: For midwestern growers, a spring flowering crop can be produced in early to mid-April from a December sowing. Space at 3–4 in. and allow for 2 to 3 sets of support. Grow on at 50° to 52° F (10° to 11° C) nights. Larkspur can be sown earlier for mid-winter cuts in the North. However, these usually lack good color and close spacing on the flower stem. Sowings made in September will flower in mid to late March. Additional lighting may help, but could be uneconomical.

Field growing: For mild-winter areas, sow mid-September to October for blooming plants during March. Larkspur has shown itself to be hardy when sown in the fall in the coastal areas of California and south of the 38th parallel (Washington, D.C.; to St. Louis; to San Francisco). It can be sown nearly year-round in California, except in the early summer months when heat is the limiting factor. For midwestern and northeastern states, direct sowings should be made in April and May due to the warm weather. For

Ball Culture: There is no crop for which so much cultural information has accumulated during the past ten years as this one. Eustoma is native to Texas and has been valued for its overall performance as a cut flower or as a pot plant. Over 50 years ago, the Japanese started breeding Eustoma to perfect its cut flower performance and since the early 70s this crop has been a staple in the Japanese cut flower trade. As these varieties were perfected, additional ones were introduced into the market and many of these were then, in turn, introduced into the American trade. A number of cut flower growers as well as the Walt Disney World gardeners significantly boosted Eustoma's popularity among growers and home gardeners. During the 1980s, varieties in new shades and bicolors were introduced, as were many with double flowers. In the '90s, pot plant varieties have emerged; new selections will be made more available as the years progress.

In regards to culture and how to grow these plants, keep in mind that they are a long term crop and proper attention to their specific requirements must be met. As a cut flower, sowings can be made in early February; for flowering plants the third to fourth week of June. With the range of varieties on the market today, the latest varieties may flower up to 3 weeks later than the earliest varieties, with the doubles often requiring the most time to flower.

The key to a successful crop is keeping the plant from rosetting. Stress brought on by cool temperatures can stall a plant's development from a seedling to a flowering plant. Most prevalent during the first 7 weeks of development, but especially so during the first 4 weeks. Once the flats are removed from the germination area, grow on at 65° F nights and 78° F days.

If plants have rosetted they will need a chilling period of 50° F between 4 and 5 weeks to reverse this process. In the southern U.S., plug producers have used intermittent lighting at night to reverse the rosetted plants instead.

Individual seedlings of Eustoma can be transplanted, though this can add to the rosetting problem already described if the roots are damaged. It will take approximately 1½ months to develop to the third or fourth true leaf stage before the plants are big enough to handle for transplanting. The seedlings often have small leaves and so will be tiny at the time of transplanting. Remember that the small leaves on the top of the media are hiding a massive root system, often 8 to 10 times greater than the aboveground growth. It is difficult to get the whole root; damage leads to rosetted plants. Either be sure to leave plenty of root space for this crop to develop or else buy in or produce your own plugs.

Once the fifth set of true leaves develops on the plants, there is less chance of rosetting due to the stresses of light and temperature. Temperatures can be lowered to 60°–65° F at night and 70°–75° F during the day. Lower temperatures can be used but often produce weaker plants, while higher temperatures can initiate root diseases brought on by Fusarium, Rhizoctonia, or Botrutos.

Flower buds of many varieties are initiated under long days. Research suggests that a number of varieties bred for winter production (flowering under short days) are weaker and less aesthetically pleasing than other ones. The Recommended Varieties sections lists varieties which have been grown successfully and addresses the available selections in more detail.

LEGEND • LIGHTING: L–Light   C–Cover   L/C–Light or Cover   C. LT.–Cover Lightly

| Green Packs | Crop Time (Weeks) | | | | No. Plants | | Pinching | Spacing | | Staking | | Field Height | Seed Required | |
	Flower Packs	Pots 4-in./10-cm.	Baskets 10-in./25.5-cm.		Pot 4-in./10-cm.	Basket 10-in./25.5-cm.		Ghse.	Field	Ghse.	Field		Ghse. For 1,000 Plants	Field For 1,000 Sq. Ft./ 90-Sq. M.
—	—	—	—		—	—	No	3x4 in. 8x10 cm.	*	2–3 tiers	1 tier	3–4 ft. .9–1.2 m.	1/4 oz. 7.1 g.	2–3.5 oz. 56.8–99.4 g.

summer-blooming plants in the Midwest, sow seed direct to the field in May for blooming plants 10 to 13 weeks later, depending on the weather.

*Since larkspur is sown direct to the field, spacings must be adjusted to suit your own planting equipment. As a guideline, use approximately 4 to 6 in. (10 to 15 cm.) between plants.

Green Packs	Flower Packs	Pots 4-in./10-cm.	Baskets 10-in./25.5-cm.	Pot 4-in./10-cm.	Basket 10-in./25.5-cm.	Pinching	Ghse.	Field	Ghse.	Field	Field Height	Ghse. For 1,000 Plants	Field For 1,000 Sq. Ft./ 90-Sq. M.
—	—	20–24**	—	1–2	—	*	4x6 in. 10x15 cm.	6x6 in. 15x15 cm.	1 tier	1 tier	2–2.5 ft. 61–76 cm.	1/256 oz. 110 mg.	—

Recommended varieties: For cut flower production, there are too many selections available to cover in this short segment. The following represents some of the readily available selections found in the U.S. trade. Heidi is a variety I prefer. It has a tulip-like bloom and keeps its petals upright even as it ages. Available in 9 separate colors as well as a mixture, the variety is a spray type in which the center bloom colors up first, followed shortly by a number of secondary flowers. Yodel is a larger flowering variety with cup-shaped blooms similar to a morning glory bloom. But, though larger, they often fall apart faster than does Heidi. Yodels have been used in the cut flower trade but may have more merit as a bedding plant or tall pot plant instead. The Royal series of Eustoma are about the same height as the Yodels with a flower shaped in between Yodel and Heidi. Flowers are larger than Heidi and Royal Purple is a vibrant color that has no comparison. Finally, the Echo series is a double flowering variety available in 8 separate colors as well as a mixture. The variety is promoted as 100% double flowering but may produce some semi-doubles under stress.

There are several pot plant selections at the time of this writing. Mermaid Blue is a deep blue flowering variety that often blooms two weeks behind Blue Lisa, and while similar in color, Blue Lisa tends to be an inch or two taller.

Both varieties can be grown in 4-in. pots but pots appear fuller if you grow 3 plants per 5-in. pot. Little Bell Blue is the latest of the three to flower (about 3 to 5 days later than Mermaid Blue). It is the tallest of the three varieties, though growers have used B-Nine to keep the plants dwarf.

Most varieties sold on the U.S. trade are pelleted for easier handling.

*Eustoma is commonly grown without a pinch.

**To grow Eustoma as a pot plant, sow seeds in early February to flower during June. However, a Florida grown 10-week-old plug (220 tray) transplanted with 1 plug per 4 in. pot during the last week of March started flowering 10 weeks later. By comparison, a 32-cell pack was planted from plugs at the same time and started flowering 8 weeks after transplanting with 50% color one week later. An interesting note: plants for use in the landscape had an overall better performance if they were grown in a 4 in. pot and transplanted with bud but limited color. Those planted from cell packs in full bloom stalled when transplanted to the display in early June and only recovered when blooms were picked off and ample water was given. Those transplanted from pots performed well all season until frost. Be sure to space plants close together since Eustoma varieties grow erect with limited branching at their base.

CUT FLOWERS

Class	Number of Seeds	Germination Temperature		Lighting	Days To Germinate	Days Sowing To Transplant	Growing On Temperature	
		Fahrenheit	Celsius				Fahrenheit	Celsius
MARIGOLD, AFRICAN **Tagetes erecta** *Also see Bedding, Florist and Foliage Plant section.*	9,000/oz. 315/g.	72°–75°	22°–24°	C. Lt.	7	10–15	65°–68°	18°–20°
NIGELLA **N. damascena**	11,000/oz. 385/g.	65°–70°	18°–21°	L/C	7–14	*	60°–62°	15°–17°
SAPONARIA **S. vaccaria**	6,400/oz. 229/g.	68°–72°	20°–22°	C	4–8	*	65°	18°
SCABIOSA–Annual **S. atropurpurea** *Also see Perennial Plants section.*	6,000/oz. 210/g.	65°–70°	18°–21°	C	10–12	20–29	50°–55°	10°–13°

Ball Culture: The African marigold is the strain used most often for cut flowers since the French types tend to be too short and the triploid varieties are too expensive. A native American plant, cut marigolds sell surprisingly well as wildflower bouquets for restaurants. If the scent of the foliage is overpowering, it can be removed to increase the blooms' sales potential.

Greenhouse growing: In the Chicago area, August sowings will not flower freely until late December or January, and only one cutting can be made. Plants should be grown on a raised bench at 55° to 58° F (13° to 14° C) without excess moisture to prevent Botrytis. If grown too cool, marigolds will stop growing altogether. Sown in early February, plants will flower in time for a late Easter.

Field growing: This is the most common growing method for cut flower production of marigolds. Seed can be sown direct to the field or sown in the greenhouse and transplanted after 4 to 5 weeks. African marigolds flower in 15 to 18 weeks, depending on crop times for individual varieties and whether the terminal bud is removed to create a spray effect.

Ball Culture: Nigella does not respond well to transplanting, so sowing direct to the final bed or field once the danger of frost has passed is recommended. Germination rates can be quite low–often less than 50%–so order seed accordingly.

Greenhouse growing: Due to the nature of this crop, greenhouse growing isn't recommended. However, growers who would like to try this method should sow in small pots and transplant into a raised bed or bench. Sow in the fall or early spring for flowering crops approximately 3 months later.

Field growing: In areas along the 38th parallel (Washington, D.C.; St. Louis; San Francisco) and south, nigella can be sown direct to the field in the fall, just as you would larkspur. Plants bloom during the spring of the following year. Sow seed where it is to flower, and space tightly to allow each seedling to produce 1 terminal flower. Generally, a spring sowing to the field takes 9 to 12 weeks to flower.

Ball Culture: S. vaccaria is an annual flowering plant which blooms quickly, sets seed and then dies. Sow seed directly where it is to flower in the field. Sowings made from the last frost date until June will yield blooming plants in 6 to 7

weeks with full bloom expected within 10 days after the start of flowering. Harvest the whole plant for use as a cut flower and remember that these plants will not flower again. Treat saponaria as you would annual gypsophila and sow seed

every 10 days in the spring to have color over several weeks.

For decades, there have only been two varieties available. Pink Beauty is a medium-rose-pink color that does not fade.

Ball Culture: Scabiosas are cool-weather tolerant, performing best in night temperatures of 50° to 55° F (10° to 13° C) in the greenhouse, or 5° to 8° F (3° to 4.5° C) warmer in the field. Germination may be irregular.

Greenhouse growing: The primary problem with greenhouse growing is the weak-stemmed cuts that are usually produced by the warm temperatures (65° F/18° C and higher). However, some good cuts have been produced from a mid-February sowing. Benched in April, these will flower in mid to late May. Keep night temperatures at 50° to 55° F (10° to 13° C), and regulate the daytime temperatures to keep the greenhouse as close to 60° to 62° F (16° to 17° C) as possible. During the winter, this crop needs an additional 6 hours of light to promote elongation and flowering.

Field growing: This is the recommended method for growing scabiosas. They can be sown direct or started in the greenhouse and transplanted to the field from 2¼ or 3-in. (6 to 7.5-cm.) pots. In the Midwest, sow seed in March and transplant in May. Field sowing can be done anytime after the ground warms up until hot weather begins. In warm winter areas, seed can be planted direct to the field in the fall for spring flowering. These flower stems will be of higher quality than those from a spring sowing.

CUT FLOWERS

Green Packs	Crop Time (Weeks)			No. Plants		Pinching	Spacing		Staking		Field Height	Seed Required	
	Flower Packs	Pots 4-in./10-cm.	Baskets 10-in./25.5-cm.	Pot 4-in./10-cm.	Basket 10-in./25.5-cm.		Ghse.	Field	Ghse.	Field		Ghse. For 1,000 Plants	Field For 1,000 Sq. Ft./90-Sq. M.
9–10	11–12	12–13	—	1	—	No	8x10 in. 20x25.5 cm.	10x10 in. 25.5x25.5 cm.	Yes	No	30 in. 76 cm.	1/8–1/4 oz. 3.6–7.1 g.	2–3 oz. 56.8–85.2 g.

Recommended varieties: The Gold Coins and Jubilees are the best series for cut flower use; Gold Coins offers the earliest flowering and strongest show of color.

| — | — | — | — | — | — | No | 6x8 in. 15x20 cm. | 6x6 in. 15x15 cm. | No | No | 2–2.5 ft. 61–76 cm. | 1/4 oz. 7.1 g. | 3–3.5 oz. 85.2–99.4 g. |

The entire plant is pulled as soon as it blooms. The pods may be used for fresh cuts or for dried arrangements once the flowers have died. Nigella's fresh cut flowers have elicited little commercial interest, but local markets have had better response. Blue-flowered nigellas are the most preferred.

*The best cuts of nigella come from the first blooms, so we recommend successive sowings several weeks apart to extend the harvesting season.

| — | — | — | — | — | — | No | — | 2x4 ft. 61x1.22 m. | No | No | 18–24 in. 46–61 cm. | 1/4 oz. 7.1 g. | .2–.3 oz. 6-9 g. |

Flowers measure 3/4 in. across and the plants are very free-flowering. The white-flowering counterpart to Pink Beauty is White Beauty, which meets the same criteria as pink but with a pure white flower instead. The flowers have no scent.

*Do not transplant individual seedlings of saponaria since the resulting plants will be weak and short-lived and bear limited flower color.

| 10–12 | — | — | — | 1 | — | No | 8x8 in. 20x20 cm. | 12x24 in. 30.5x61 cm. | 1 tier | 1 tier | 2.5–3 ft. 76–91.5 cm. | 1/4–1/2 oz. 7.1–14.2 g. | 3–3.5 oz. 85.2–99.4 g. |

Both perennial and annual forms of scabiosa are used for cut flower production. The flower colors in perennial varieties are limited to shades of blue and white, while annual varieties also include pink, salmon, scarlet, deep rose, lavender and many other colors. Both annual and perennial varieties are often short-lived here in the Midwest due to our hot summers.

CUT FLOWERS

Class	Number of Seeds	Germination Temperature		Lighting	Days To Germinate	Days Sowing To Transplant	Growing On Temperature	
		Fahrenheit	Celsius				Fahrenheit	Celsius
SNAPDRAGON Antirrhinum majus *Also see Bedding, Florist and Foliage Plants section.*	180,000/oz. 6,300/g.	70°–75°	21°–24°	L	7–14	15–20	45°–50°	7°–10°
STATICE Limonium sinuata *Also see Bedding, Florist and Foliage Plants and Perennials sections.*	8,500/oz.* 297/g.*	70°	21°	L/C	5–12	12–19	50°–55°	10°–13°
Limonium perezii	24,000/oz.* 840/g.*	70°	21°	L/C	5–12	12–19	50°–55°	10°–13°
Phylliostachys suworowii	195,000/oz.* 6,825/g.*	70°	21°	L/C	7–15	21–28	55°–58°	13°–14°

Ball Culture: Greenhouse-grown cut snapdragons produce longer stems and, in many cases, larger individual florets than their garden counterparts. Greenhouse cut snaps are a prime crop to grow in a cool greenhouse for fall and winter production in the northern U.S..

Greenhouse cut snaps are divided into 4 groups based on their response to light, temperature and daylength. The varieties within Group 1 respond well to low light, low temperature and short daylength while Group 4 responds to the opposite–high light, long days, and higher temperatures than those in Group 1.

In general, a mid- to late-August sowing will flower mid-December to February depending on temperatures; plants grown at 55° to 58° F will flower sooner than those grown at 45° to 48° F. An early September sowing will start blooming in February, while an early-December sowing will flower in May.

Ball Culture–Limonium sinuata: This category includes varieties like the Fortress, Excellent and Turbo series. To grow as a bedding plant, see the Bedding, Florist and Foliage Plants section. For sowing and germination, the seed can be left exposed or covered lightly, as indicated. Some growers cover lightly with vermiculite to increase the humidity around the seed and provide more uniform germination.

For greenhouse production as a cut flower in the Midwest and Northeast, sow seed in January and transplant to the bench in March or April for flowering starting in May. Remember to keep the growing temperatures at 50° to 55° F (10° to 13° C) after the plants are established in the pack or bench. When these temperatures are maintained for 4 to 5 weeks, sinuata varieties will flower earlier than when grown warmer. Our experience has shown that fall sowings in the Midwest produce weak-stemmed plants, stretched from the low light conditions. (L. sinuata is not photoperiodic but flowers freely under long days.)

Sowings earlier than January may not be profitable in the Midwest when grown as greenhouse cuts.

For field growing in the Midwest, 8 to 10-week old plants can be transplanted to the field from the time the danger of frost is past until June. Once established, the plants will flower until frost. In warm winter areas, successive sowings in the greenhouse every 6 to 8 weeks will provide continuous cropping from December until June or later, depending on the climate. Plants are grown in packs or as plugs, and planted to the field instead of sowing seed direct. In areas where the temperatures seldom drop below 55° F (13° C), some growers have used gibberillic acid in the field at 500 ppm when plants are 6 to 8 in. (15 to 20 cm.) across in order to promote earlier flowering. However, the application has produced mixed results.

Harvest cuts when the blooms fold outward and show color.

As a fresh cut, flowers last 1 to 2 weeks and will dry naturally in the arrangement. Stored fresh at 34° to 36° F (1° to 2° C), sinuatas will last 2 to 3 weeks. For the best quality stems, use blooms the first or second week. To dry statice, hang a bunch of 7 to 10 stems upside down in a dark, well-ventilated area for 2 weeks.

Ball Culture–Limonium perezii: Commonly called seafoam statice, L. perezii is often sold as a perennial in the Midwest, but is short-lived at best. This crop does best in areas with mild winters, such as coastal California, where the variety will live from year to year and grow to 3 ft. (91.5 cm.) tall. Sowings in mid-March produce green packs for sale by June 1. Sowings made before mid-March and transplanted to the field once the danger of frost has passed tend to flower earlier. Though plants bloom the same season from seed, we recommend the other 2 varieties to Midwest growers rather than L. perezii.

Ball Culture–L. suworowii: Also called rat-tail or Russian statice, suworowii types were favored in northern greenhouses as fresh cut flowers at the height of the cut flower era, long before bedding plants gained popularity.

L. suworowii can be used either dried or fresh.

For greenhouse production in the Midwest and Northeast, sow seed in September for harvesting in February. Either grow as plugs or transplant from the seedling tray into 2 1/4-in. (6-cm.) cell packs. L. suworowii can also be transplanted directly into the final bed. Grow on at 55° F (13° C) nights.

CUT FLOWERS

| Green Packs | Crop Time (Weeks) | | | No. Plants | | Pinching | Spacing | | Staking | | Field Height | Seed Required | |
	Flower Packs	Pots 4-in./10-cm.	Baskets 10-in./25.5-cm.	Pot 4-in./10-cm.	Basket 10-in./25.5-cm.		Ghse.	Field	Ghse.	Field		Ghse. For 1,000 Plants	Field For 1,000 Sq. Ft./ 90-Sq. M.
—	—	—	—	—	—	No	5x5 in. 13x13 cm.	6x6 in. 15x15 cm.	2 tiers	No*	**	1/128 oz. 221 mg.	—

For more complete scheduling information, refer to the Ball Seed Catalog.

* Forcing snapdragons are most often grown in the greenhouse where they reach 4 to 6 ft. in height. For field production, green-house varieties can be used, but both they and garden snap selections like Rocket should be harvested as first or second color shows. Unfortunately, bees pollinate the flowers and the blooms drop shortly thereafter.

** Field comments are based only on Rocket snapdragons and not the greenhouse forcing varieties.

Green Packs	Flower Packs	Pots 4-in./10-cm.	Baskets 10-in./25.5-cm.	Pot 4-in./10-cm.	Basket 10-in./25.5-cm.	Pinching	Ghse.	Field	Ghse.	Field	Field Height	Ghse. For 1,000 Plants	Field For 1,000 Sq. Ft./ 90-Sq. M.
8–10	**	—	—	—	—	No	10x10 in. 25.5x25.5 cm.	12x12 in. 30.5x30.5 cm.	No	No	18–24 in. 46–61 cm.	1/4 oz. 7.1 g.	—
8–10	**	—	—	—	—	No	10x10 in. 25.5x25.5 cm.	12x12 in. 30.5x30.5 cm.	No	No	24–36 in. .6–.9 m.	1/16–1/8 oz. 1.8–3.6 g.	—
9–11	**	—	—	—	—	No	8x8 in. 20x20 cm.	10x10 in. 25.5x25.5 cm.	No	No	15–20 in. 38–51 cm.	1/128–1/64 221–443 mg.	—

For field production of suworowii, growers on the West Coast and in the deep South should sow at the same time or a little earlier, and transplant to the field from pots or as plugs. In dark winter areas, try growing this crop in 5 to 6-in. (13 to 15-cm.) pots to flower. As plants become root-bound, they tend to flower more readily. Incandescent lights also help.

*The approximate seed per ounce/gram listed above refers to cleaned seed only, and not seed heads.

**Statice becomes too large in the pack to sell in bloom. We recommend selling plants green or with early color. L. perezii varieties require more time than other types to flower and should be grown and sold in 4-in. (10-cm.) or larger pots for spring sales. In the Midwest, L. perezii does not reach the size and free-flowering performance that it does in mild winter areas.

CUT FLOWERS

Class	Number of Seeds	Germination Temperature		Lighting	Days To Germinate	Days Sowing To Transplant	Growing On Temperature	
		Fahrenheit	Celsius				Fahrenheit	Celsius
STOCK **Matthiola incana** *Also see Bedding, Florist and Foliage Plants section.*	19,000/oz. 630/g.	65°–75°	18°–24°	L/C	7–14	11–18	50°–55°	10°–13°

Ball Culture: At the Chicago latitude, stock can be sown from July 15 to February 15 for flowering from January to early June. Sowings at other times may be blind. Grow plants on at 50° to 55° F (10° to 13° C) nights once they are established. Stock varieties are available in either single or double-flower forms; the market commands more money for double flowering types. Unfortunately, the seed of both comes in the same packet and cannot be separated easily among the Giant Column or Miracle selections.

Stock varieties are divided up into dwarf and cut flower types available in selectable versus nonselectable types.

Basically, selectable varieties are those in which single- and double-flowering forms can be distinguished by their leaf color and shape. Single-flowering varieties will have a smooth leaf edge and darker green foliage than their double-flowering counterparts.

Due to their cost, selectable varieties are almost always greenhouse grown and often require a short vernalization treatment of between 50° to 58° F for 3 to 4 weeks. They are far more expensive then nonselectable types, which are most often field grown varieties from either the Giant Column or Miracle series.

Class	Number of Seeds	Germination Temperature		Lighting	Days To Germinate	Days Sowing To Transplant	Growing On Temperature	
XERANTHEMUM **X. annuum** **(Immortelle)**	20,000/oz. 700/g.	70°–72°	21°–22°	L/C	10–15	18–27	60°	15°

Ball Culture: Xeranthemum can be either sown direct to the field or transplanted out. However, it is our experience that taller, more free-flowering crops are produced on transplanted crops started as plugs. It appears that transplanting individual seedlings, though common, may stress the plant enough to eventually decrease the quality of the cut flower. The effect of transplanting is less noticeable when they are used as bedding plants.

Field growing: Seed sown direct to the field in April and May will produce flowering plants from mid to late July through September.

Xeranthemum is harvested for use as a dried cut flower as soon as the blooms show full color.

Class	Number of Seeds	Germination Temperature		Lighting	Days To Germinate	Days Sowing To Transplant	Growing On Temperature	
ZINNIA **Z. elegans**	2,000–6,000/oz. 70–210/g.	70°–72°	21°–22°	C	3–7	10–15	60°	15°

Ball Culture: Field growing: Sow direct to the field once the soil warms, until early July. In the Chicago area, sowings made in early June have flowered by late July, with a final sowing in early July flowering just before frost.

Greenhouse growing: This method isn't generally used since many varieties have high light requirements. Without the proper light, hollow stems tend to develop which cannot support the weight of the flowers and they fall over.

If you disbud plants–remove all side buds, leaving only the terminal–you will produce higher quality cut flowers.

Recommended varieties: Among the tall varieties, State Fair and Ruffles are the most popular choices for cut flowers. Both are available in separate colors, but State Fair is usually grown as a mixture and added to other summer flowering cuts in mixed bouquets.

| Green Packs | Crop Time (Weeks) | | | No. Plants | | Pinching | Spacing | | Staking | | Field Height | Seed Required | |
	Flower Packs	Pots 4-in./10-cm.	Baskets 10-in./25.5-cm.	Pot 4-in./10-cm.	Basket 10-in./25.5-cm.		Ghse.	Field	Ghse.	Field		Ghse. For 1,000 Plants	Field For 1,000 Sq. Ft./ 90-Sq. M.
—	—	—	—	—	—	No	3x6 in. 7.5x15 cm.	6x8 in. 15x20 cm.	*	No	2.5–3 ft. 76–91.5 cm.	1/4–1/2 oz. 7.1–14.2 g.	1–1.5 oz. 28.4–42.6 g.
—	—	—	—	—	—	No	—	8x8 in. 20x20 cm.	—	No	18–24 in. 46–61 cm.	1/8 oz. 3.6 g.	3.5 oz. 99.4 g
5–6	8–9	9	—	1	—	No	8x8 in. 20x20 cm.	12x12 in. 30.5x30.5 cm.	Yes	No	26–32 in. 66–81 cm.	3/4–1 oz. 21.3–28.4 g.	2.5–3.5 oz. 71–99.4 g.

Greenhouse growing: Sowings made in early August, benched 2 to 4 weeks after sowing, will flower from late January to mid-February. A mid-December sowing will flower in mid-May.

Field growing: In the midwestern and eastern U.S., we recommend treating stocks as a greenhouse crop. For field growing, the best areas include the California coastal region and southern Arizona.

*Support material is not necessary if grown at the night temperatures indicated.

CUT FLOWERS

Class	Number of Seeds	Seed for 1,000 plants (oz.)	Germination Temperature		Lighting	Days To Germinate	Days Sowing To Transplant	Growing On Temperature	
			Fahrenheit	Celsius				Fahrenheit	Celsius
ACANTHUS A. mollis (Bear's Breech)	130/oz. 5/g.	10	68°–72°	20°–22°	C	15–25	22–40	60°	15°

Ball Culture: A. mollis is an excellent accent plant for use in the garden. Though it is not perennial in severe winters, it provides deep glossy green foliage throughout its life cycle in our northern summer gardens.

A. mollis is a large-seeded item that germinates readily. Old seed should be soaked overnight in warm tap water prior to sowing. Sow seed into a deep container to allow for its quickly-developing roots to grow down. A mid-February sowing can be transplanted to A-18's or 3-in. pots by mid-to late-March but will probably have to be potted up again into 6-in. or gallon containers within 2 to 4 weeks as daytime light and temperatures increase.

Class	Number of Seeds	Seed for 1,000 plants (oz.)	Germination Temperature		Lighting	Days To Germinate	Days Sowing To Transplant	Growing On Temperature	
ACHILLEA A. filipendulina (Fernleaf Yarrow)	200,000/oz. 7,000/g.	1/128–1/64	65°–70°	18°–21°	L	10–15	18–24	50°	10°
A. millefolium (Common Yarrow, Milfoil)	140,000/oz. 4,900/g.	1/64	65°–70°	18°–21°	L	10–15	18–24	50°	10°
A. ptarmica (Sneezeweed)	90,000/oz. 3,150/g.	1/64–1/32	65°–70°	18°–21°	L	10–15	18–24	50°	10°
A. tomentosa cv. Aurea (Woolly Yarrow)	175,000/oz. 6,200/g.	1/128	65°–70°	18°–21°	L	4–8	15–25	50°	10°

Ball Culture: Grown at night temperature of 55° F (13° C), February and March sowings will flower sporadically the same season. Of the species noted here, A. filipendulina is the slowest to flower along with A. tomentosa cv. Aurea, while A. ptarmica flowers readily from seed the same year.

Most achilleas produce the strongest show of color from June to July. The majority of seed-propagated millefolium varieties bloom in pastel colors that readily fade in full sun, though Summer Pastels will hold its color better throughout the season.

Yarrows are best used as fresh or dried cut flowers; harvest as soon as flowers open and show color.

The leaves of A. tomentosa cv. Aurea are covered with a fine layer of white hairs from which spikes of muted gold-

Class	Number of Seeds	Seed for 1,000 plants (oz.)	Germination Temperature		Lighting	Days To Germinate	Days Sowing To Transplant	Growing On Temperature	
ALYSSUM Aurinia saxatilis Compacta (Basket of Gold)	30,000/oz. 1,050/g.	1/16	60°–70°	15°–21°	L	7–14	18–24	50°	10°

Ball Culture: Seed can be either sown direct to the final container or transplanted from a sowing tray after germination. For green packs, have plants ready for sale along with pansies and annual alyssum varieties. Perennial alyssums

do not flower the same year from seed. Sow in June and July for quart sales the following spring.

These perennials perform better if established in the garden or pot before the onset of warm weather. Alyssums are excellent for rock gardens and growing over walls, displaying attractive, gray-green foliage.

Class	Number of Seeds	Seed for 1,000 plants (oz.)	Germination Temperature		Lighting	Days To Germinate	Days Sowing To Transplant	Growing On Temperature	
ANAPHALIS A. margaritacea (Pearly Everlasting)	600,000/oz. 21,000/g.	1/256	65°–68°	18°–20°	L/C	4–8	21–28	55°–60°	10°–15°

Ball Culture: Hardy perennials in the Midwest, these gray-green plants are ideal for mass plantings or cut flowers. For fresh or dried cuts, harvest plants when the yellow centers

of blooms become visible. Pearly Everlastings are drought tolerant and can become invasive once established.

PERENNIALS

Crop Time (Weeks) Green Packs	Number of Plants QT./.95 L.	GAL./3.8 L.	For Same Year Flowering	Blooming Months	Hardiness Zones	Garden Height	Garden Spacing	Staking	Location
n/a	1	1	1A	June–July	6–10	2–4 ft. .61–1.2 m.	15–24 in. 38-61 cm.	No	F. Sun to P. Shade

A 6-in. pot with one plant per container will be salable by early to mid May from a mid-February sowing. Plants will then be up to 5-6 in. tall in the pot when transplanted to the garden.

Related material: In our northern winters we have better luck over wintering A. spinosissimus (formerly A. spinosus), which is commonly called spiny bear's breeches. This plant can be dangerous to grow due to the rigid white spines on each of its deeply-lobed leaves. It is hardy to zone 5 but may require some over-wintering-protection. It can be propagated by seed or by divisions taken in early fall or spring. Beware when trying to divide these plants—they are unforgiving if you get too close.

Crop Time (Weeks) Green Packs	Number of Plants QT./.95 L.	GAL./3.8 L.	For Same Year Flowering	Blooming Months	Hardiness Zones	Garden Height	Garden Spacing	Staking	Location
10–12	1	1	1A, 1C, 2	June–Aug.	3–8	4 ft. 1.2 m.	15–18 in. 38–46 cm.	No	F. Sun
10–12	1	1	1A, 1C, 2	June–Aug.	3–8	24–30 in. 61–76 cm.	12–15 in. 30.5–38 cm.	No	F. Sun
10–12	1	1	1C, 2	June–Aug.	3–8	20–28 in. 51–71 cm.	10–12 in. 25.5–30.5 cm.	No	F. Sun
8–10	1–2	1–2	1A, 2	June–July	3-8	1–1.5 ft. 30–45 cm.	12–15 in. 30–37.5 cm.	No	F. Sun to P. Shade

en-yellow flowers develop. Beware of planting in areas with frequent overhead watering or misting, which increases the chance of Botrytis, especially during extended periods of high humidity.

Sowings made in February will produce salable green packs (32 cells per flat) in about 9 weeks at 50° F night temperatures. Leaves of A. tomentosa cv. Aurea are deep green when young but acquire the woolly appearance as spring progresses and leaves mature. When planted to the garden in May, plants will flower sporadically the first season but color will improve the following year. This culture also fits the other Achilleas listed here.

Crop Time (Weeks) Green Packs	Number of Plants QT./.95 L.	GAL./3.8 L.	For Same Year Flowering	Blooming Months	Hardiness Zones	Garden Height	Garden Spacing	Staking	Location
10–12	1–2	2–3	1A, 1C	April–May	3–7	10–12 in. 25.5–30.5 cm.	10–12 in. 25.5–30.5 cm.	No	F. Sun to P. Shade
10–11	1–2	1–2	1A	July–Aug.	3–8	24–26 in. 61–66 cm.	12 in. 30.5 cm.	No	F. Sun to P. Shade

Class	Number of Seeds	Seed for 1,000 plants (oz.)	Germination Temperature Fahrenheit	Celsius	Lighting	Days To Germinate	Days Sowing To Transplant	Growing On Temperature Fahrenheit	Celsius
ANTHEMIS A. tinctoria cv. Kelwayi	60,000/oz. 2,100/g.	1/32	70°–72°	21°–22°	C. Lt.	3–7	12–18	60°	15°

Ball Culture: Anthemis grows quickly from seed and often looks weedy in the flat or pot. Once it has been transplanted and the roots have established themselves in the pack (10 to 15 days after transplanting), gradually reduce the night temperature to 45° to 50° F and the day temperature by 5 to 8 degrees.

Anthemis doesn't hold well in the pack. Once it's ready to sell, it needs to be moved within 10 days or it will appear weedy and uneven. However, this quick development gives

Class	Number of Seeds	Seed for 1,000 plants (oz.)	Germination Temperature Fahrenheit	Celsius	Lighting	Days To Germinate	Days Sowing To Transplant	Growing On Temperature Fahrenheit	Celsius
AQUILEGIA A. caerulea (Columbine)	15,000– 22,000/oz. 525–770/g.	1/8	70°–75°	21°–24°	L	21–28	30–40	50°–55°	10°–13°

Ball Culture: Aquilegia seed should be chilled 2 to 3 weeks at 40° F (4° C) before sowing to improve uniformity. During germination, reduce soil temperature at night by 5° to 8° F (3° to 4° C).

For especially troublesome germination, or if seed is being kept from year to year, try stratifying seed in moistened sand in a refrigerator for several weeks to break dormancy. An application of fungicide on each sowing tray after seed-

ing has also been noted to improve germination by as much as 10-15%. In our trials, we have achieved May flowering from sowings made in early October, transplanted to packs in early November, potted in February, and grown at

Class	Number of Seeds	Seed for 1,000 plants (oz.)	Germination Temperature Fahrenheit	Celsius	Lighting	Days To Germinate	Days Sowing To Transplant	Growing On Temperature Fahrenheit	Celsius
ARABIS A. alpina (Rock Cress)	70,000/oz. 2,450/g.	1/32	65°–70°	18°–21°	L	6–12	14–21	50°–55°	10°–13°

Ball Culture: Sow from June to August for blooming plants the following year. Grow cold during winter months in 4-in. (10-cm.) pots or quart (.95 l.) containers.

Locate only in well-drained areas of the garden with afternoon shade.

Class	Number of Seeds	Seed for 1,000 plants (oz.)	Germination Temperature Fahrenheit	Celsius	Lighting	Days To Germinate	Days Sowing To Transplant	Growing On Temperature Fahrenheit	Celsius
ARMERIA A. maritima (Thrift, Sea-Pinks)	20,000/oz. 714/g.	1/8	68°–70°	20°–21°	L/C	4–10	15–22	50°–55°	10°–13°

Ball Culture: Soak seed in warm water overnight before sowing. Armerias are very slow to grow. February sowings grown at 55° to 62° F (13° to 17° C) nights hardly fill out the pack by early May, much less a 4-in. (10-cm.) pot.

Sow in winter for green packs in spring, allowing 12 to 15 weeks at night temperatures of 62° to 65° F (17° to 18° C)

Class	Number of Seeds	Seed for 1,000 plants (oz.)	Germination Temperature Fahrenheit	Celsius	Lighting	Days To Germinate	Days Sowing To Transplant	Growing On Temperature Fahrenheit	Celsius
ASCLEPIAS A. tuberosa (Butterfly Weed)	3,500/oz. 122/g.	1/2	70°–75°	21°–24°	L/C	21–28	35–55	60°–65°	15°–18°

Ball Culture: If germination proves difficult, chill seed in moistened sand at 36° to 40° F (2° to 4° C) for several weeks prior to germination.

Sowings made from December to March will flower from May to July of the same year. If possible, transplant into deep cell-packs or 4-in. (10-cm.) pots to allow development of large taproots. Asclepias plants grow quickly and do not

need cold treatment to flower. Over-wintered in cold frames, plants require more time to grow and develop as they emerge in spring.

Class	Number of Seeds	Seed for 1,000 plants (oz.)	Germination Temperature Fahrenheit	Celsius	Lighting	Days To Germinate	Days Sowing To Transplant	Growing On Temperature Fahrenheit	Celsius
ASTER—Perennial A. alpinus (Michaelmas Daisy) *Also see Bedding, Florist and Foliage Plants, and Cut Flowers sections.*	24,000/oz. 840/g.	1/16	65°–70°	18°–21°	L	14–21	21–30	55°–60°	13°–15°

Ball Culture: Chill seed between sowings. They have a longer crop time than the fall-flowering types. Grown warm at 65° F (18° C), plants can be finished in under 18 weeks, but may need a growth regulator to maintain good shaping.

PERENNIALS

Crop Time (Weeks) Green Packs	Number of Plants		For Same Year Flowering	Blooming Months	Hardiness Zones	Garden Height	Garden Spacing	Staking	Location
	QT./.95 L.	GAL./3.8 L.							
9–10	1	1–2	1A	June–July	3–7	26–36 in. 66–91.5 cm.	12–15 in. 30.5–38 cm.	No	F. Sun

rise to blooming plants in the garden in late June from a 10- week transplant placed in the garden in early May.

Plants will flower from seed the same year, and will continue to flower all summer long.

15–20	1	1–2	1A, 1C, 3	May–June	3–8	18–30 in. 46–76 cm.	12–15 in. 30.5–38 cm.	No	P. Shade

40°F (4° C). Allow 15 to 20 weeks for green pack sales, depending on the variety.

This perennial is most often sold in 4-in. (10-cm.) pots or quart (.95 l.) containers, which requires starting plants the year prior to sale to allow ample growth time. Aquilegias

have an open habit and need 12-in. (30.5 cm.) spacing to fill in well.

12–15	1	1–3	1A, 1C	April–May	3–7	6–8 in. 15–20 cm.	10 in. 25.5 cm.	No	P. Shade

Planted in full shade in our trials, these perennials have proven to be short-lived, lasting 3 years at best.

14–17	1–2	2–3	1A, 1C	May–June	4–8	6–15 in. 15–28 cm.	10 in. 25.5 cm.	No	F. Sun

to keep plants actively growing. Flowering cannot be expected the first season, through some blooming may

occur sporadically.

11–13	1–2	2–3	2, 1A	June–Aug.	4–9	18–24 in. 46–61 cm.	10–12 in. 25.5–30.5 cm.	No	F. Sun

Sell green in packs or 4-in. (10-cm.) pots along with warm-season annuals in the spring. A proven garden performer, this warm-season perennial loses all its foliage when cool

weather sets in. Asclepias plants can also be used as fresh cut flowers by searing the stems.

The dried seed pods are excellent for fall decorating.

18–22	1	1–2	1A, 1C	May–June	4–9	10–12 in. 25.5–30.5 cm.	10–12 in. 25.5–30.5 cm.	No	P. Shade

PERENNIALS

Class	Number of Seeds	Seed for 1,000 plants (oz.)	Germination Temperature Fahrenheit	Celsius	Lighting	Days To Germinate	Days Sowing To Transplant	Growing On Temperature Fahrenheit	Celsius
ASTILBE A. x Arendsii (False Spirea)	384,000/oz. 13,440/g.	1/256	60°–70°	15°–21°	L	14–21	40–50	60°–65°	15°–18°

Ball Culture: From seed, astilbes are particularly slow to grow and develop. Sowings made from February to mid-March and grown at 60° F (15° C) nights will be ready for green pack sales in mid-June. Plants will not flower the same season. Sowings made in June or July and established in quart (.95 l.) containers before winter can be

Class	Number of Seeds	Seed for 1,000 plants (oz.)	Germination Temperature Fahrenheit	Celsius	Lighting	Days To Germinate	Days Sowing To Transplant	Growing On Temperature Fahrenheit	Celsius
AUBRIETA A. deltoidea (False Rockcress)	85,000/oz. 2,975/g.	1/64–1/32	65°–70°	18°–21°	L	14–21	20–25	50°–55°	10°–13°

Ball Culture: There are differing opinions on the best germinating temperatures for this crop. Some authorities recommend 60° F (15° C), and note that germination rates decrease at 65° or 70° F (18° or 21° C). In our trials, 70% germination has been obtained from summer sowings germinated at 70° F (21° C) without a covering. Seed sown in January for May sales will not bloom until the following spring. Sow in summer for March sales, keeping night temperatures cool, approximately 60° F (15° C).

Class	Number of Seeds	Seed for 1,000 plants (oz.)	Germination Temperature Fahrenheit	Celsius	Lighting	Days To Germinate	Days Sowing To Transplant	Growing On Temperature Fahrenheit	Celsius
BELLIS B. perennis (English Daisy)	140,000/oz. 4,900/g.	1/64	70°–75°	21°–24°	L	7–14	15–22	50°–55°	10°–13°

Ball Culture: Grow English daisies just like pansies. December sowings will flower in spring, grown at night temperatures of 50° F (10° C). Small-flowered varieties such as Pomponette Mix tend to flower earlier than the larger types. English daisies are best treated as annuals in hot summer locations; plants do not tolerate hot weather and die by the end of summer if not grown cool.

Class	Number of Seeds	Seed for 1,000 plants (oz.)	Germination Temperature Fahrenheit	Celsius	Lighting	Days To Germinate	Days Sowing To Transplant	Growing On Temperature Fahrenheit	Celsius
BERGENIA B. cordifolia (Heartleaf Bergenia)	110,000/oz. 3,850/g.	1/64	70°–75°	21°–24°	L	4–8	45–63	55°–60°	13°–15°

Ball Culture: If germination is too low, seed can be sown into flats in December and kept at 55° to 60° F (13° to 15°C) for 2 to 3 days. Sowing flats should then be transferred to a cold frame or placed under a protected covering and layered with snow. Leave at 32° to 41° F (0° to 5° C) for 6 to 8 weeks, then bring flats back into the greenhouse to complete germination. An easier, though less dependable, way to improve germination is to chill seed for several weeks between 32° to 41° F (0° to 5° C) before sowing.

Class	Number of Seeds	Seed for 1,000 plants (oz.)	Germination Temperature Fahrenheit	Celsius	Lighting	Days To Germinate	Days Sowing To Transplant	Growing On Temperature Fahrenheit	Celsius
CAMPANULA C. var. calycanthema (Cup and Saucer, Canterbury Bells)	79,000/oz. 2,765/g.	1/64–1/32	70°	21°	C. Lt.	14–21	20–30	55°–60°	13°–15°
C. carpatica (Carpathian Harebells)	200,000/oz. 7,000/g.	1/128	70°	21°	C. Lt.	14–21	20–30	55°–60°	13°–15°

Ball Culture: After sowing, lightly cover seed to prevent dehydration; in our trials, this has also increased germination by 10% to 15%. Exposing seed to alternate light/dark periods has been known to improve germination rates as well. C. calycanthema is a true biennial which needs a cold period to flower, and should be grown at 50° F (10° C) nights. These varieties will not bloom the same year from late winter or spring sowings. The carpaticas, however, are warm-season perennials which are dependably hardy in Midwest climates and flower year after year. To grow these campanulas, sow seed in early March. Transplant in late March and plant in late May for flowering plants by late July.

Class	Number of Seeds	Seed for 1,000 plants (oz.)	Germination Temperature Fahrenheit	Celsius	Lighting	Days To Germinate	Days Sowing To Transplant	Growing On Temperature Fahrenheit	Celsius
CARNATION Dianthus carophyllus *Also see Bedding, Florist and Foliage Plants section.*	14,000/oz. 490/g.	1/8	65°–70°	18°–21°	C. Lt.	5–13	15–20	50°–55°	10°–13°

Ball Culture: Carnations' spicy, clove-scented blooms are held well above the slender, gray-green leaves. A half-strength dose of fungicide when sowing has increased germination rates by 10-15% in our tests. Started early enough, plants grown at 50° F (10° C) will flower the same season. Carnations respond well to Cycocel.

Crop Time (Weeks) Green Packs	Number of Plants		For Same Year Flowering	Blooming Months	Hardiness Zones	Garden Height	Garden Spacing	Staking	Location
	QT./.95 L.	GAL./3.8 L.							
14–16	1	1	1A, 1C	July–Aug.	4–8	24–36 in. 61–91.5 cm.	10–12 in. 25.5–30.5 cm.	No	F. Sun to P. Shade

over-wintered for sales the following spring. These plants will flower the same season they are sold.

Crop Time (Weeks) Green Packs	Number of Plants		For Same Year Flowering	Blooming Months	Hardiness Zones	Garden Height	Garden Spacing	Staking	Location
10–12	1	2–3	1A, 1C	April–May	4–7	6–8 in. 15–20 cm.	8–10 in. 20–25.5 cm.	No	P. Shade

When sowing in the fall, drop temperatures to between 50° and 55° F (10° and 13° C) once seedlings are established in the final quart (.95 l.) container or 4-in. (10-cm.) pot. Late

September sowings grown at 50° F (10° C) nights will produce blooming plants by early May. Aubrietas are

small enough to grow in 32 cell-per-flat packs; 72-cell flats restrict plant growth.

Crop Time (Weeks) Green Packs	Number of Plants		For Same Year Flowering	Blooming Months	Hardiness Zones	Garden Height	Garden Spacing	Staking	Location
14–15*	1	1	2	May–June	4–8	6–8 in. 15–20 cm.	10–12 in. 25.5–30.5 cm.	No	F. Sun to P. Shade

We have sown Pomponette Mix in early October, transplanted in mid-October and produced flowering plants by late February, grown at 45° F (10° C).

*Crop time is based on spring blooming of Super Enorma Mix. Pomponette Mix flowers 2-3 weeks earlier.

Crop Time (Weeks) Green Packs	Number of Plants		For Same Year Flowering	Blooming Months	Hardiness Zones	Garden Height	Garden Spacing	Staking	Location
15–18	1	1	1A, 1C	April–May	3–8	15–18 in. 38–46 cm.	12–15 in. 30.5–38 cm.	No	F. Sun to P. Shade

Mid-winter sowings will not flower the same season. February and March sowings will be ready for spring sales in large packs (18 per flat) or 4-in. (10-cm.) pots from June

to late July. For sales the following spring, sow in June and establish in quart (.95 l.) containers to over-winter in a cold frame. In the garden, bergenias display glossy, showy leaves

which can be used as cut foliage in floral arrangements.

Crop Time (Weeks) Green Packs	Number of Plants		For Same Year Flowering	Blooming Months	Hardiness Zones	Garden Height	Garden Spacing	Staking	Location
13–15	1	1–3	1B	June	3–8	26–30 in. 66–76 cm.	10–12 in. 25.5–30.5 cm.	No	F. Sun
11–12	1	1–2	2, 1A, 1C	July–Aug.	3–8	12–15 in. 30.5–38 cm.	12 in. 30.5 cm.	No	F. Sun to P. Shade

C. carpatica varieties make excellent 4-in. pot plants sold around Mother's Day for eventual planting into the flower garden as a perennial. Both the Clips and Uniform cultivars are available in either white or blue flower colors (mixtures

are available too) and will flower reliably 15 to 18 weeks after sowing for May sales. Plants can be grown for sales earlier in the year but require additional lighting to tone the plants up and increase flower bud count. In University and

grower tests, C. carpatica is responsive to Cycocel applications.

Crop Time (Weeks) Green Packs	Number of Plants		For Same Year Flowering	Blooming Months	Hardiness Zones	Garden Height	Garden Spacing	Staking	Location
10–12	1–2	2–3	1B, 2	June–July	5–8	18–20 in. 46–51 cm.	12–15 in. 30.5–38 cm.	Yes	F. Sun to P. Shade

In the garden, plants are weak-stemmed and must be staked to keep the flowers upright. Our trials have proven carnations

to be short-lived, requiring a mulch to over-winter.

PERENNIALS

Class	Number of Seeds	Seed for 1,000 plants (oz.)	Germination Temperature		Lighting	Days To Germinate	Days Sowing To Transplant	Growing On Temperature	
			Fahrenheit	Celsius				Fahrenheit	Celsius
CATANANCHE C. caerulea	12,000/oz. 441/g.	1/8	65°–70°	18°–21°	L/C	4–10	15–25	60°	15°
CENTAUREA C. montana **(Perennial Cornflower, Perennial Bachelor Buttons, Mountain Bluet)**	2,200/oz. 79/g.	1	72°	22°	C. Lt.	7–14	14–21	55°–60°	13°–15°
CENTRANTHUS C. ruber **(Red Valerian, Kentranthus)**	17,500/oz. 612/g.	1/8	60°–65°	15°–18°	L/C	14–21	18–28	60°	15°
CERASTIUM C. tomentosum **(Snow-in-Summer)**	80,000/oz. 2,800/g.	1/64–1/32	65°	18°	L	7–14	15–23	55°–60°	13°–15°
CHRYSANTHEMUM—**Perennial** **C. x morifolium, now called Dendranthemum** **(Mum)**	70,000/oz. 2,450/g.	1/32	60°–70°	15°–21°	L	5–10	15–25	55°–60°	13°–15°
C. parthenium **(Matricaria)**	200,000/oz. 7,000/g.	1/64	70°	21°	L	7–10	16–26	55°–60°	13°–15°
C. coccineum* **(Pyrethrum, Painted Daisy)**	17,000–35,000/oz. 595–1,225/g.	1/16	60°–70°	15°–21°	L/C	14–21	20–35	55°–60°	13°–15°

Ball Culture: Catananche forms a rosette of leaves from which 24-in. flower stalks emerge. The light-to medium-blue flowers have a papery calyx and are often used as a dried cut flower.

In a 32-cell tray, allow up to 11-12 weeks for green pack sales from a mid-March sowing. Plants will set some minor flower

Ball Culture: C. montana flowers sporadically the same year from seed, but does better the second season.

Even then, plants bloom in cycles, flowering again several weeks after the first flush.

Once the first flush of flowering is over, centaureas tend to fall over from the center, exposing the crowns. This is where new shoots develop.

Ball Culture: These easy-to-grow perennials flower the same year from seed. Sow in late February, transplant in March and sell after all danger of frost has passed for blooming plants in June.

*This crop can also be sold in flowering packs, allowing 16 to 18 weeks for crop time; however, plants are usually too

Ball Culture: Displaying eye-catching, gray-green foliage, cerastiums do best in sunny locations; plants tend to stretch and be short-lived in full shade.

Ball Culture—C. x morifolium: This easily-grown crop is best treated as an annual or short-lived perennial. Sow seed in March for pack sales in late May and early June. Perennial chrysanthemums bloom under short days in late summer. In the Midwest, plants are hardy, but don't last long.

C. parthenium: Matricarias are best treated as annuals since few plants make it through the winter; re-seeding accounts for most of the plants which appear the following spring. Early March sowings, transplanted to packs in late March, have flowered in the field by late June in our trials. Plants get tall, so they are best sold green in packs.

Recommended varieties: Under **C. x morifolium**, Autumn Glory and the Fashion series are ideal for pot and pack production. The Fashions are available in separate colors, with some color variation within each shade.

Of the **C. partheniums**, Ball Double White Improved is a popular

Crop Time (Weeks) Green Packs	Number of Plants		For Same Year Flowering	Blooming Months	Hardiness Zones	Garden Height	Garden Spacing	Staking	Location
	QT./.95 L.	GAL./3.8 L.							
11–13	1–2	1–2	2,1A	July–August	4–8	18–24 in. 45–60 cm.	10 in. 25.5 cm.	No	P. Shade

buds by the end of July with additional buds appearing until frost. The plants often flower more freely the second year.

10–12	1	1	2, 1A, 1C	May–Sept.	3–8	15–24 in. 38–61 cm.	12–15 in. 30.5–38 cm.	No	F. Sun

Dormant buds form on side branches and flower 4 to 6 weeks later. The number of secondary blooms is limited by staking unless plants are cut back. Plants reseed themselves when flowers are allowed to mature and seeds develop.

8–10*	1	1–2	2	June–Sept.	5–9	24–36 in. 61–91.5 cm.	12–15 in. 30.5–38 cm.	No	F. Sun

tall if allowed to flower in the pack.

11–12	1–2	2–3	1A, 1C	May–June	3–7	6–8 in. 15–20 cm.	10–12 in. 25.5–30.5 cm.	No	F. Sun to P. Shade
10–11	1	1	2	Sept.–Oct.	5–9	10–12 in. 25.5–30.5 cm.	10–12 in. 25.5–30.5 cm.	No	F. Sun
10–12	1	1	2	July–Aug.	5–8	10–24 in. 25.5–61 cm.	10 in. 25.5 cm.	No	F. Sun
10–12	1	1–2	2, 1A	May–June	4–9	20–24 in. 51–61 cm.	10–12 in. 25.5–30.5 cm.	No	F. Sun

standard used for cutting and the back of borders. This matricaria is very reliable, producing 100% double blooms from seed.

Santana is one of our most popular 10-in. (25.5-cm.) dwarfs, performing best in full sun as a border plant. Another 10-in. (25.5-cm.) dwarf, Golden Ball, features ball-shaped, golden yellow flowers.

Note: Seed varieties of fall flowering mums are not as reliable or true-to-type as their vegetative counterparts.

C. coccineum: Sown in January, plants will flower in July of the same year. Pyrethrums produce a stronger show of color the following season.

*Botanical names have been changed several times; pyrethrums may also be listed as Chrysanthemum roseum, P. coccineum or P. roseum.

Class	Number of Seeds	Seed for 1,000 plants (oz.)	Germination Temperature Fahrenheit	Germination Temperature Celsius	Lighting	Days To Germinate	Days Sowing To Transplant	Growing On Temperature Fahrenheit	Growing On Temperature Celsius
COREOPSIS C. grandiflorum (Tickseed)	10,000/oz. 350/g.	1/4	65°–75°	18°–24°	L	9–12	20–25	55°	13°
DAISY, SHASTA Chrysanthemum superbum*	15,000– 35,000/oz. 525–1,225g.	1/16	65°–70°	18°–21°	C. Lt.	9–12	20–25	55°–60°	13°–15°
DELPHINIUM D. grandiflorum var. chinense	29,000/oz. 1,015/g.	1/16	—	—	C	12–18	20–28	50°–55°	10°–13°
D. x cultorum	8,000– 10,000/oz. 280–350/g.	1/4	—	—	C	12–18	20–28	50°–55°	10°–13°
D. x belladonna	8,000– 10,000/oz. 280–350/g.	1/4	—	—	C	12–18	20–28	50°–55°	10°–13°

COREOPSIS

Ball Culture: This is one of the easiest perennials to grow and displays wide variation between varieties.

Standing 20 in. (51 cm.) tall, Sunray has fully-double, golden yellow blooms up to 2 in. (5 cm.) across, and offers a more uniform performance than its cohort, Sunburst.

However, Sunburst provides a more vigorous appearance, growing 30 to 35 in. (76 to 91.5 cm.) tall with 2-in. (5-cm.) semi-double flowers of golden yellow.

Sowings made before May have been known to lead to a fall flowering and, in some cases, winter-kill. Sowing in summer for fall sales or over-wintering in pots for sale the following spring may increase the plants' life span.

DAISY, SHASTA

Ball Culture: Most often treated as a biennial in the central U.S., shasta daisies need well-drained areas to avoid winter-kill. Even then, Midwest winters are generally too severe for plants to last more than 2 or 3 seasons.

Crop times and flowering response vary widely between varieties. Alaska is the standard for single-flowered varieties. Grown cool at 50° to 55° F (10° to 13° C) under group 2 guidelines, it will flower in July and August from a February

sowing and reach 2 ft. (6 1 cm.) in height.

A 12 to 15-in. (30.5 to 38-cm.) dwarf single, Silver Princess should be grown under 1A and/or 1C guidelines since plants do not flower dependably the same season from seed. We have sown this crop in early October, then transplanted to packs in late October and to pots in early February. Grown at 48° to 50° F (9° to 10° C), plants flowered in late June; earlier flowering can be forced by raising temperatures.

Among the tallest of the singles, Starburst reaches 36 to 48 in. (.9 to 1.2 m.) in height which makes it ideal for cutting. This variety has not displayed the hardiness of Alaska in our trials. Treat Starburst under group 1A and/or 1C guidelines; plants will not flower the same season from seed.

With a height of only 10 in. (25.5 cm.), Snow Lady is the shortest dwarf single in our line. For green 4-in. (10-cm.) pot sales in May and June, sow in January and February; for

DELPHINIUM

Ball Culture: Germination is the key to success with this crop.

Germination temperatures vary among varieties. For the Giant Pacific Court series, alternate day/night temperatures of 80° F (26° C) and 70° F (21° C) work well. Other varieties require temperatures of 65° to 75° F (18° to 24° C) to germinate.

Delphiniums go into dormancy after harvesting. Studies have shown that an application of gibberellic acid can help

to break the cycle; however, this information is intended only as a research update and is not a recommendation. Delphiniums are best treated as annuals in the southern U.S. since they prefer cool climates.

Packs are not recommended for growing delphiniums since, like zinnias, they are susceptible to root rot brought on by overwatering.

We recommend pot production only. January sowings grown at 50° to 55° F (10° to 13° C) nights along with

other bedding plants will be ready for green sales in quart (.95 l.) containers in the spring. Plants will flower in June of the same year. For spring sales of gallon (3.8 l.) containers, 1 to 2 plants each, sow in July and August and over-winter in the final containers. Seed can also be sown direct to the ground in a cold frame during August, transplanted to gallon (3.8 l.) containers in March and April, and sold in the spring. This crop has difficulty developing plants large enough to flower in hot weather if sown after February 20. Plants must be protected against freezing in winter.

PERENNIALS

Crop Time (Weeks) Green Packs	Number of Plants		For Same Year Flowering	Blooming Months	Hardiness Zones	Garden Height	Garden Spacing	Staking	Location
	QT./.95 L.	GAL./3.8 L.							
11–12	1	1–2	*	June–July	4–9	20–36 in. 51–91.5 cm.	12–15 in. 30.5–38 cm.	No	F. Sun

Another cultivar, Early Sunrise, reaches 2 ft. (61 cm.) in height, with semi-double, golden-yellow blooms. Our trials, however, have shown that plants grow only 18 to 20 in. (46 to 51 cm.) tall in their first year. Early Sunrise will flower any time of the year. We have sown seed in late August for flowering 5–6 in. pot plants 16–18 weeks later

grown at 60° F at night. Plants were not grown directly under HID or fluorescent lights but additional lighting could help improve the overall quality of the plant. If grown with warm days, you may want to experiment with a growth regulator or drop the temperatures to keep the plant in check.

*In the first season, these varieties flower sporadically at best, and should be treated as 1A or 1C classifications.

| see Culture below | 1 | 1 | — | June–Aug. | 5–8 | see Culture below | — | see Culture below | F. Sun |

flowering pots, sow in September and October. Grown under group 2 guidelines, plants will flower the same year from seed.

Slightly taller than Snow Lady and with larger blooms, White Knight exhibits pure white flowers on plants that do well in either 4½-in. or quart containers. White Knight blooms 10–14 days later than Snow Lady.

Since it is impossible to produce 100% doubles from seed, it is best to grow double-flowered cultivars vegetatively. Both G. Marconi and Dieners Double should be treated under 1A or 1C guidelines; neither will flower the same year from seed, but both boast 3½-in. (9-cm.) flowers and 50% doubleness.

In the garden, space Snow Lady 10 in. (25.5 cm.) apart, White Knight, Alaska and Silver Princess 12 to 15 in. (30.5 to 38 cm.) apart and Starburst on 15 in. (38 cm.) centers.

Shasta daisies do best in full sun. Stake Starburst to keep flowers upright.

*Shasta daisies may be listed as C. leucanthemum or C. maximum; however, they are classified as C. superbum in *Hortus Third*.

—	1	1	2	June–July	4–7	10–24 in. 25.5–61 cm.	12–15 in. 30.5–38 cm.	No	F. Sun
—	1	1	1A, 1C, 2	June–July	4–7	2–6 ft. 61–152 cm.	12–15 in. 30.5–38 cm.	No	F. Sun
—	1	1	1A, 1C, 2	June–July	4–7	24–30 in. .9–1.8 m.	12–15 in. 30.5–38 cm.	Yes*	F. Sun

Recommended varieties: For stately columns of blooms, the Giant Pacific Court series is by far the best choice. A number of outstanding colors are available, including royal blue (King Arthur), dark blue (Black Knight) and mid-blue (Blue Bird). Plants reach 5 ft. (1.5 m.) in height with elegant double flowers featuring distinct bees of white or darker shades of the bloom color. Though Giant Pacific Courts is stronger and longer-stemmed, Belladonna, Bellamosum and Casablanca give a more free-flowering performance. An F₂ hybrid, the Century series grows as tall as the Giant Pacific Courts with an upright habit and stronger stems.

No staking has been needed for this series in our trials.

Delphinium grandiflorum var. chinense cultivars are available in blue, white and shades of color in between. The blue flowering selections are the most popular and most frequently found color on the U.S. trade. Sowings made in early January and transplanted to 4 in. pots in February will flower by mid-May when grown at 45° F.

*Only the Giant Pacifics must be staked.

Class	Number of Seeds	Seed for 1,000 plants (oz.)	Germination Temperature		Lighting	Days To Germinate	Days Sowing To Transplant	Growing On Temperature	
			Fahrenheit	Celsius				Fahrenheit	Celsius
DIANTHUS D. barbatus (Sweet William)	25,000/oz. 875/g.	1/16	60°–70°	15°–21°	C. Lt.	7–10	13–18	55°–58°	13°–14°
D. deltoides (Maiden Pinks)	48,000/oz. 1,680/g.	1/32	60°–70°	15°–21°	C. Lt.	7–10	13–18	55°–58°	13°–14°
D. plumarius (Cottage Pinks)	22,000/oz. 770/g.	1/16–1/8	60°–70°	15°–21°	C. Lt.	7–10	13–18	55°–58°	13°–14°

Ball Culture: Sow early to mid-February and grow on at 50° to 55° F (10° to 13° C) for green pack sales in April and May. Plants will flower the following year. For flowering spring sales, sow the previous summer, then transplant to quart (.95 l.) or gallon (3.8 l.) containers and over-winter plants in a cold frame or under protective covering.

Biennial varieties of Dianthus barbatus are available, such as Double Midget Mix, Double Mix and Indian Carpet Mix.

Class	Number of Seeds	Seed for 1,000 plants (oz.)	Germination Temperature		Lighting	Days To Germinate	Days Sowing To Transplant	Growing On Temperature	
DIGITALIS D. purpurea (Foxglove)	126,000/oz. 4,410/g.	1/64	60°–65°	15°–18°	L	5–10	15–20	55°–60°	13°–15°

Ball Culture: Sow this biennial in late February or March for green quart sales in April.

Excelsior Hybrid is the standard Foxglove, the one most commonly known by consumers. A warm blend of several different colors, this mixture does well when the culture guidelines are followed. Our other cultivar, Foxy, is a brighter mixture that can be forced into flower in 5 months.

Sown in December or January, Foxy will bloom in May and June. Do not grow in cell packs other than to size the plants

Class	Number of Seeds	Seed for 1,000 plants (oz.)	Germination Temperature		Lighting	Days To Germinate	Days Sowing To Transplant	Growing On Temperature	
DORONICUM D. cordatum (Leopard's Bane)	26,000/oz. 910/g.	1/16	70°	21°	L	14–21	25–36	50°	10°

Ball Culture: Germination for doronicums is always low; order seed accordingly. To our knowledge, there is no special treatment available to improve germination percentages.

This crop has shown itself to be a truly hardy perennial in our Midwest trials. Grown in full sun, however, plants have experienced problems with the heat and humidity, particularly in August.

Class	Number of Seeds	Seed for 1,000 plants (oz.)	Germination Temperature		Lighting	Days To Germinate	Days Sowing To Transplant	Growing On Temperature	
ECHINACEA PURPUREA* (Purple Coneflower)	8,000/oz. 280/g.	1/4	70°	21°	L/C	5–10	20–28	50°	10°

Magnus is an attractive, rose-red purpurea cultivar that flowers the same year from seed if started early, in January or February. Growing 30 to 40 in. (76 to 101.5 cm.) tall, Magnus offers both a uniform habit and consistently-colored, 4 to 5-in. (10 to 13-cm.) blooms.

Class	Number of Seeds	Seed for 1,000 plants (oz.)	Germination Temperature		Lighting	Days To Germinate	Days Sowing To Transplant	Growing On Temperature	
ECHINOPS E. ritro (Globe Thistle)	2,600/oz. 91/g.	1	65°–72°	18°–22°	L	14–21	20–31	55°–60°	13°–15°

Ball Culture: We do not recommend echinops for packs since plants are susceptible to root rots brought on by over watering. For best performance, transplant directly to final containers–preferably 3-in. (7.5-cm) pots.

Echinops are excellent cut flowers. Harvest as the blue blooms emerge, and use stems fresh or dried.

PERENNIALS

Crop Time (Weeks) Green Packs	Number of Plants		For Same Year Flowering	Blooming Months	Hardiness Zones	Garden Height	Garden Spacing	Staking	Location
	QT./.95 L.	GAL./3.8 L.							
10–12	1	1–2	1B	May–June	4–8	6–18 in. 15–46 cm.	12 in. 30.5 cm.	No	P. Shade
10–12	1	1–2	1A, 1C	May–June	4–8	6–9 in. 15–23 cm.	12 in. 30.5 cm.	No	F. Sun to P. Shade
10–12	1	1–2	1A, 1C	May–June	4–8	12–16 in. 30.5–41 cm.	12 in. 30.5 cm.	No	F. Sun to P. Shade
see culture below	1	1	1B	May–June	4–8	36–60 in. .9–1.5 m.	15–18 in. 38–46 cm.	No	P. Shade

up before transplanting to a larger container such as a quart or gallon pot. Plants grow very quickly, and if you have the time to transplant them, their performance later could be damaged.

Crop Time (Weeks) Green Packs	Number of Plants		For Same Year Flowering	Blooming Months	Hardiness Zones	Garden Height	Garden Spacing	Staking	Location
12–14	1	1–2	1A	April–May	4–7	18–24 in. 46–61 cm.	10 in. 25.5 cm.	No	P. Shade
10–12	1	2–3	2, 1A	July–Sept.	3–9	30–40 in. 76–102 cm.	12 in. 30.5 cm.	No	F. Sun

* Sometimes referred to as Rudbeckia purpurea

Crop Time (Weeks) Green Packs	Number of Plants		For Same Year Flowering	Blooming Months	Hardiness Zones	Garden Height	Garden Spacing	Staking	Location
11–13	1	1	1A, 1C	July–Aug.	4–9	24–36 in. 61–91.5 cm.	12–15 in. 30.5–38 cm.	No	F. Sun

PERENNIALS

Class	Number of Seeds	Seed for 1,000 plants (oz.)	Germination Temperature Fahrenheit	Celsius	Lighting	Days To Germinate	Days Sowing To Transplant	Growing On Temperature Fahrenheit	Celsius
EUPHORBIA E. myrsinites (Myrtle Euphorbia)	2,500/oz. 88g.	—	65°–70°	18°–21°	L	7–14	16–22	50°–55°	10°–13°
E. polychroma (Cushion Spurge)	6,500/oz. 227/g.	3/4–1	65°–70°	18°–21°	L	8–15	20–25	50°–55°	10°–13°
GAILLARDIA G. x grandiflora (Blanket Flower)	1,700/oz. 61/g.	1	70°–75°	21°–24°	L	5–15	20–30	55°–60°	13°–15°
GAURA G. lindheimeri	1,700/oz. 61/g.	1	70°–72°	21°–22°	L	5–11	21–28	50°	10°
GEUM G. quellyon	10,000/oz. 350/g.	1/8–1/4	65°–70°	18°–21°	L	8-15	18-28	55°–60°	13°–15°
GYPSOPHILA G. repens (Creeping Baby's Breath)	45,000/oz. 1,575/g.	1/16	70°–80°	21°–26°	L	5–10	15–20	50°	10°
G. paniculata (Baby's Breath)	26,000–30,000/oz. 910–1,050/g.	1/16–1/8	70°–80°	21°–26°	L	5–10	15–20	50°	10°

Ball Culture—E. myrsinites: This is a slow-growing perennial that produces whorls of gray-green foliage during the summer. From seed, the plants often have deep-to mid-green foliage color which lightens as the plants mature in the cell pack and in the garden. Sowings made in early February and transplanted to cell packs (32 cells per flat) will be salable green in June. Plants seldom flower the same season from seed though there may be sporadic bloom. Plants will not basal branch until they are firmly established in the cell pack, and even then the secondary stems are limited. Plants will freely branch in the garden and be strong and full by the time cold weather arrives.

Ball Culture—E. polychroma: Producing only small, incon-

Ball Culture: Seed can be sown directly to the final containers, although at least one transplanting is recommended. When sown around January, plants flower sporadically the first season. Gaillardias are among the few perennials which flower all season long, but flowering performance decreases after the main flush in June.

Recommended varieties: Reaching 2 ft. (61 cm.) in height, Monarch Strain makes a good cut flower. A shorter cultivar, Goblin, blooms earlier and performs best in perennial borders.

Ball Culture: An easy-to-grow perennial from seed that bears large, white blooms on plants 3-4 ft. tall. Gaura is native to the south central U.S. and can become invasive in warm-winter areas. Sowings made in early February and transplanted to 32 cells per flat will be salable green within 9 weeks; slightly longer if growing under cooler night temperatures. If sown

Ball Culture: Decreasing temperatures by 8° to 10° F (4° to 6° C) at night helps increase germination rates, if overall stand is germinating poorly due to use of old seed.

Ball Culture: Sow G. repens in mid-February and grow on at 55° F (13° C) nights for late June flowering. In our trials of paniculata cultivars, fall sowings grown at 48° F (9° C) nights tended to produce columnar-shaped plants without breaks. Plants were slow to grow and develop, and their overall health was poor. Sowings made in February will flower sporadically the first year from seed. Gypsophilas can be used fresh or dried.

Recommended varieties: The repens types offer 2 especially good choices, Repens White and Repens Rose (light pink). Though the white offers a stronger habit, they are both ideal for borders and trailing over rock walls with a

Crop Time (Weeks) Green Packs	Number of Plants		For Same Year Flowering	Blooming Months	Hardiness Zones	Garden Height	Garden Spacing	Staking	Location
	QT./.95 L.	GAL./3.8 L.							
14–17	1	1–2	1A, 1C	April–May	5–9	8–10 in. 20–25.5 cm.	12 in. 30 cm.	No	F. Sun
8–11	1	1	1A, 1C	April–May	4–8	12–16 in. 30.5–41 cm.	12 in. 30.5 cm.	No	F. Sun

spicuous blooms, E. polychroma is sold for its crimson fall foliage.

Without pre-treatment, germination percentages often fall below 40%. If time permits, chill seed for 4 to 6 weeks at

35° to 40° F (2 ° to 4° C), then sow seed on moistened (not wet) sand. This cold sowing method is derived from that used for bergenias. If this method cannot be followed, we have found that covering the seed in sowing medium to a depth 2 to 3 times the width of the seed increases germina-

tion by 15%. Stems eventually fall over from the center, exposing the crown; within several weeks the crown branches out again.

Crop Time (Weeks) Green Packs	Number of Plants		For Same Year Flowering	Blooming Months	Hardiness Zones	Garden Height	Garden Spacing	Staking	Location
10–12	1	1–2	1A, 1C	June–Frost	3–9	14–24 in. 35.5–61 cm.	12 in. 30.5 cm.	No	F. Sun
9–10	1	1	2, 1A	June–August	5–9	3–4 ft. 91.5–122 cm.	12–15 in. 30.5-38 cm.	No	F. Sun

in March, plants will be ready in approximately the same amount of time, though flowering effect during the summer appears weaker than that of February-sown plants.

Crop Time (Weeks) Green Packs	Number of Plants		For Same Year Flowering	Blooming Months	Hardiness Zones	Garden Height	Garden Spacing	Staking	Location
10–11	1	1	1A, 1C	May–June	4–8	24 in. 61 cm.	12–15 in. 30.5–38 cm.	No	F. Sun to P. Shade
10–12	1–2	2–3	2, 1A, 1C	June–July	4–8	6 in. 15 cm.	10 in. 25.5 cm.	No	F. Sun
10–12	1–2	2	1A, 1C	June–July	4–8	30–36 in. 76–91.5 cm.	12 in. 30.5 cm.	Yes	F. Sun

height of just 6 in. (15 cm.).

Among the paniculata cultivars, there are both single and double white-bloomed types. The free-flowering singles

grow up to 30 in. (76 cm.) tall and must be staked once plants bloom. Grown from seed, Snowflake is an upright cultivar that reaches 36 in. (91.5 cm.) and displays 50% double and 50% single blooms. For 100% doubles, try

Bristol Fairy or the larger-flowered Perfecta. Both are excellent performers; however, seed is unavailable since these are vegetatively produced.

PERENNIALS

95

Class	Number of Seeds	Seed for 1,000 plants (oz.)	Germination Temperature		Lighting	Days To Germinate	Days Sowing To Transplant	Growing On Temperature	
			Fahrenheit	Celsius				Fahrenheit	Celsius
HELENIUM H. autuminale	140,000/oz. 5,000/g.	1/64	72°	22°	L	8-12	28-35	55°	13°

Ball Culture: Helenium is an easy-to-grow perennial suggested for the late summer garden. Sowings in February will be salable in cell packs by early May, though they develop better when used as one plant per 3 or 4 in. pot so that the leaves lay flat. Plants will start to flower in August and will continue blooming until mid-September.

Class	Number of Seeds	Seed for 1,000 plants (oz.)	Fahrenheit	Celsius	Lighting	Days To Germinate	Days Sowing To Transplant	Fahrenheit	Celsius
HELIANTHEMUM H. nummularium (Rockrose)	14,000/oz. 490/g.	1/8	70°–75°	21°–24°	L/C	15–20	25–38	50°–55°	10°–13°

Ball Culture: Growing low to the ground, helianthemums have an almost shrub-like appearance. Once established in the bed, the bases of stems become woody. These perennials have lived for a number of years in our gardens.

Class	Number of Seeds	Seed for 1,000 plants (oz.)	Fahrenheit	Celsius	Lighting	Days To Germinate	Days Sowing To Transplant	Fahrenheit	Celsius
HELIOPSIS H. helianthoides (Oxeye Daisy)	8,000/oz. 280/g.	1/4	65°–70°	18°–21°	L/C	3–10	11–20	60°	15°

Ball Culture: These plants have proven themselves to be hardy perennials in our gardens. Sown in the spring, plants show color in June of the same year and will bloom more profusely once established. Oxeye daisies are an excellent choice for Midwest gardens, flowering right up until frost.

Class	Number of Seeds	Seed for 1,000 plants (oz.)	Fahrenheit	Celsius	Lighting	Days To Germinate	Days Sowing To Transplant	Fahrenheit	Celsius
HESPERIS H. matronalis (Sweet Rocket)	14,000/oz. 490/g.	1/8	70°–80°	21°–26°	L	5–7	14–20	55°	13°

Ball Culture: Hesperis is not necessarily a true perennial; mid-March sowings for spring pack sales have only rosetted in the garden and then flowered the second year. Though listed for hardiness in zones 3 to 8, we have found this crop to be short-lived at best. Plants reseed freely.

Class	Number of Seeds	Seed for 1,000 plants (oz.)	Fahrenheit	Celsius	Lighting	Days To Germinate	Days Sowing To Transplant	Fahrenheit	Celsius
HEUCHERA H. sanguinea (Coral Bells)	500,000 oz. 175,000/g.	1/256	65°–70°	18°–21°	L	21–30	30–45	55°	13°
H. micrantha (Alumroot, Palace Purple)	500,000/oz. 175,000/g.	1/256	65°–70°	18°–21°	L	21–30	35–55	55°	13°

Ball Culture: Slow-growing perennials, heucheras produce very small seedlings much like begonias. As a result, transplanting takes more time than with most crops.

Recommended varieties: The 2 most commonly grown cultivar, Splendens and Bressingham Hybrids, are sanguinea types. Though similar in height, Splendens produces rose-red flower spikes while Bressingham Hybrids is a bright mixture of scarlet, rose and deep pink.

Purple Palace is sold for its foliage rather than its flowers. It offers an unusual combination of deep bronze foliage and yellow blooms. When growing this crop, less than 15% green leaves will appear at germination. Leave seedlings exposed to light for leaves to color up before transplanting. Crop time for Purple Palace is 2 weeks longer than for sanguinea types.

Crop Time (Weeks) Green Packs	Number of Plants		For Same Year Flowering	Blooming Months	Hardiness Zones	Garden Height	Garden Spacing	Staking	Location
	QT./.95 L.	GAL./3.8 L.							
9–11	1	1	1A, 1C	July–September	3–8	4–5 ft. 1.2–1.5 m.	12 in. 30.5 cm.	Yes	F. Sun
10–12	1	1–2	1A	June	5–9	8–10 in. 20–25.5 cm.	10 in. 25.5 cm.	No	P. Shade
8–10	1	1	2, 1A, 1C	June–Sept.	3–9	36 in. 91.5 cm.	12–15 in. 30.5–38 cm.	Yes	F. Sun
11–13	1	1–2	1B	June	3–8	24 in. 61 cm.	12 in. 30.5 cm.	Yes	F. Sun to P. Shade
8–10	1	1	1A	June–July	3–8	18–24 in. 46–61 cm.	12 in. 30.5 cm.	No	F. Sun to P. Shade
10–13	1	1	1A	June–July	3–8	18–24 in. 46–61 cm.	12 in. 30.5 cm.	No	F. Sun to P. Shade

Class	Number of Seeds	Seed for 1,000 plants (oz.)	Germination Temperature		Lighting	Days To Germinate	Days Sowing To Transplant	Growing On Temperature	
			Fahrenheit	Celsius				Fahrenheit	Celsius
HIBISCUS H. moscheutos (Rose Mallow)	2,100/oz. 73/g.	2–2.5	70°–80°	21°–26°	C	7–10	16–22	60°–65°	15°–18°
HOLLYHOCK Alcea rosea	3,000–6,000/oz. 105–210/g.	1/2	68°–70°	20°–21°	C. Lt.	14–21	20–30	55°–60°	13°–15°
HYPERICUM H. calycinum (Rose of Sharon)	26,000/oz. 910/g.	1/8	60°–70°	15°–21°	C. Lt.	10–21	20–32	50°	10°
IBERIS I. sempervirens (Hardy Candytuft)	10,000/oz. 350/g.	1/8–1/4	60°–65°	15°–18°	L	14–21	25–38	50°	10°
LATHYRUS L. splendens (Perennial Sweet Pea)	625/oz. 22/g.	3	55°–65°	13°–18°	C	10–20	*	50°–55°	10°–13°
LAVANDULA L. angustifolia var. Munstead	31,000/oz. 1,085/g.	1/8	65°–75°	18°–24°	L	14–21	20–32	58°–60°	14°–15°

Ball Culture: We have found that germination is highest when seed is covered lightly and a fungicide is applied when sowing. Soaking seed overnight before sowing also improves germination.

Packs are not recommended unless you are using large-cell packs, such as 18s. If plants become root-bound, they lose their lower leaves. For best results, sell in 4-in. (10-cm.) pots and quart (.95 l.) containers or larger.

Recommended varieties: Most often grown in the South, both Southern Belle and Dixie Belle have shown themselves to be hardy enough for Chicago latitudes. However, a

Ball Culture: If started early enough, hollyhocks will flower sporadically the same year. Plants do better, though, when sown the previous year for sales in the spring. Be sure to

transplant directly to final containers before tap roots form. The only exceptions are annual-flowering varieties which bloom the same year from a February sowing.

*Hollyhocks need staking only in windy areas.

Ball Culture: An evergreen foliage plant, hypericum becomes woody with age. This crop is not dependably hardy at Chicago latitudes but performs better in the southern U.S.

Ball Culture: A dependably hardy evergreen at Chicago latitudes, this perennial will not flower the same season from seed.

Ball Culture: Density of the seed coat causes germination rates to vary widely. To increase germination, nick the coat or rub with sandpaper, then soak seed overnight.

*Using 3 to 4 seeds per pot, sow directly to final containers. Sweet peas resent transplanting, so they do best in Jiffy-pots or similar containers that can be planted.

Recommended varieties: Keep in mind that these selections are all without fragrance. White and rose to pink shades are the predominant flower colors.

Ball Culture: Chill seed at 36° to 40° F (2° to 4° C) for 4 to 6 weeks before sowing to improve germination. In our experience, lavendulas germinate better when seed is left

exposed to light rather than covered with medium. This perennial has a long crop time, so grow in 4-in. (10-cm.) pots or quart (.95 l.) containers very early in the year, or

late in the previous year, for spring sales.

Crop Time (Weeks) Green Packs	Number of Plants		For Same Year Flowering	Blooming Months	Hardiness Zones	Garden Height	Garden Spacing	Staking	Location
	QT./.95 L.	GAL./3.8 L.							
—	1	1	2	July–frost	4–9	24–60 in. 61–152 cm.	15 in. 38 cm.	No	F. Sun

particularly rainy fall can induce winter-kill. Blooms last only one day.

Perennial varieties make decent 6-in. pots but be sure to

use at least 2 or 3 plants since blooms are so short lived. Sowings in early March will flower in early June without pinching or added lights. If the plants get away from you, Cycocel is registered for use on this crop to keep the height

in check. This culture has been used successfully on the Disco Belle series. It is not suggested that this method be used on Southern Belle due to its height.

Crop Time (Weeks) Green Packs	Number of Plants		For Same Year Flowering	Blooming Months	Hardiness Zones	Garden Height	Garden Spacing	Staking	Location
10–12	1	1	1B	July–Aug.	3–7	4–8 ft. 1.2–2.4 m.	15–20 in. 38–51 cm.	*	F. Sun
12–14	1	1	1A, 1C	May–June	5–8	12 in. 30.5 cm.	10–12 in. 25.5–30.5 cm.	No	F. Sun to P. Shade
10–14	1	1	1A, 1C, 3	April–May	3–8	9–12 in. 23–30.5 cm.	10–12 in. 25.5–30.5 cm.	No	F. Sun
8–10*	—	—	2	June–Aug.	4–7**	5 ft.*** 1.5 m.	12–15 in. 30.5–38 cm.	Yes	F. Sun to P. Shade

**A severe winter can kill them here in Zone 5.

*** Plants trail 5 ft., not growing erect.

Crop Time (Weeks) Green Packs	Number of Plants		For Same Year Flowering	Blooming Months	Hardiness Zones	Garden Height	Garden Spacing	Staking	Location
15–18	1	2–3	2, 1A	June–Aug.	5–9	12 in. 30.5 cm.	12 in. 30.5 cm.	No	F. Sun

Class	Number of Seeds	Seed for 1,000 plants (oz.)	Germination Temperature		Lighting	Days To Germinate	Days Sowing To Transplant	Growing On Temperature	
			Fahrenheit	Celsius				Fahrenheit	Celsius
LEONTOPODIUM L. alpinum (Edelweiss)	310,000/oz. 10,850/g.	1/128	68°–72°	20°–22°	L	15–25	30–48	50°–55°	10°–13°

Ball Culture: To increase germination rates, chill seed at 40° F (4° C) for 3 weeks before sowing. Then reduce germinating temperatures by 10° F (6° C) each night until seedlings emerge.

* Plants usually die out from winter kill in the North or extreme summer heat in the South. Plants do best in cool, coastal areas, which limits their value anywhere but coastal California.

Class	Number of Seeds	Seed for 1,000 plants (oz.)	Germination Temperature		Lighting	Days To Germinate	Days Sowing To Transplant	Growing On Temperature	
LIATRIS L. pycnostachya	9,400/oz. 329/g.	1/4	65°–70°	18°–21°	L	21–28	28–38	50°	10°
L. spicata (Kansas Gay Feather, Blazing Star)	9,400/oz. 329/g.	1/4	65°–70°	18°–21°	L	21–28	28–38	50°	10°

Ball Culture: Reduce night temperatures by 10° F (6° C) during germination. In the pack, plants appear similar to grass seedlings. This crop usually blooms sporadically the first seasons from a mid-winter sowing; however, we have had March sowings fail to flower until the following year in our trials. Unlike most flowering spike plants, these bloom from the top down.

Class	Number of Seeds	Seed for 1,000 plants (oz.)	Germination Temperature		Lighting	Days To Germinate	Days Sowing To Transplant	Growing On Temperature	
LINUM L. perenne (Flax)	9,000–24,000/oz. 315–840/g.	1/8	65°–75°	18°–24°	L/C	10–12	22–34	58°–60°	14°–15°

Ball Culture: If germination is low, drop night temperatures by 10° F (6° C) for the first 10 to 15 days. Growing in packs for spring sales, we have had excellent results using small-cell packs (such as 72s) and selling plants green. January or February sowings grown at no less then 55° F (13° C), then shifted to 4-in. (10-cm.) or quart (.95 l.) containers, will flower the same season from seed.

Linums have an open habit which requires a number of

Class	Number of Seeds	Seed for 1,000 plants (oz.)	Germination Temperature		Lighting	Days To Germinate	Days Sowing To Transplant	Growing On Temperature	
LUNARIA L. annua (Honesty Plant, Money Plant)	1,500/oz. 52/g.	1	65°–75°	18°–24°	L	10–14	16–25	55°–60°	13°–15°

Ball Culture: Easy to germinate and grow, these biennials flower the second season from seed. Once the seed pods fall off, the white, papery-looking bracts are exposed. Flowers can be dried and used in arrangements.

Class	Number of Seeds	Seed for 1,000 plants (oz.)	Germination Temperature		Lighting	Days To Germinate	Days Sowing To Transplant	Growing On Temperature	
LUPINE Lupinus polyphyllus	1,000/oz. 35/g.	2	65°–75°	18°–24°	C	6-12	18-30	50°–55°	10°–13°

Ball Culture: Soak seed for 24 hours before sowing. Plants can become fairly large by the time they are ready to sell, so use 4-in. (10-cm.) or larger pots and allow plenty of space for growing.

*If lupines must be produced in packs, allow 10 to 12 weeks and use large-cell packs (18s or 32s). Sell green; plants will flower the following year.

Crop Time (Weeks) Green Packs	Number of Plants		For Same Year Flowering	Blooming Months	Hardiness Zones	Garden Height	Garden Spacing	Staking	Location
	QT./.95 L.	GAL./3.8 L.							
10–12	1	1–3	1A, 1C	June–July	4–7*	8 in. 20 cm.	8–10 in. 20–25.5 cm.	No	F. Sun
12–14	1	1–2	1A, 1C, 2	July–Aug.	3–8	3–5 ft. .9–1.5 m.	10–12 in. 25.5–30.5 cm.	No	F. Sun
12–14	1	1–2	1A, 1C, 2	July–Aug.	3–8	18–24 in. 46–61 cm.	10–12 in. 25.5–30.5 cm.	No	F. Sun
13–15	2–3	3	2, 1A	June–Aug.	5–9	12–24 in. 30.5–61 cm.	10 in. 25.5 cm.	No	F. Sun

plants to fill out a quart (.95 l.) or gallon (3.8l) container.
Plants are short-lived at the Chicago latitude, providing only
2 to 3 years of flowering with some losses each season.

10–12	1	2–3	1B	April–June	3–8	24–26 in. 61–66 cm.	10–12 in. 25.5–30.5 cm.	No	F. Sun
*	1	1	1B	June–July	4–8	20–60 in. 51–152 cm.	15 in. 38 cm.	No	F. Sun to P. Shade

Class	Number of Seeds	Seed for 1,000 plants (oz.)	Germination Temperature		Lighting	Days To Germinate	Days Sowing To Transplant	Growing On Temperature	
			Fahrenheit	Celsius				Fahrenheit	Celsius
LYCHNIS **L. chalcedonica** **(Maltese Cross)**	50,000/oz. 1,750/g.	1/32	70°	21°	L/C	—	22–35	50°	10°
L. haageana	65,000/oz. 2,275/g.	1/32	70°	21°	L/C	—	22–35	50°	10°

Ball Culture: Chill seed at 35° F (2° C) when storing.

Both types of lychnis will flower the same season they are sown if started early in the year. Plants sown after

March 15 do not perform as strongly in the garden as those from earlier sowings.

Class	Number of Seeds	Seed for 1,000 plants (oz.)	Germination Temperature		Lighting	Days To Germinate	Days Sowing To Transplant	Growing On Temperature	
MALVA **M. moschata** **(Mallow, Musk Mallow)**	12,500/oz. 441/g.	1/8	70°–72°	21°–22°	C	3–6	14–21	50°	10°

Ball Culture: Sowings made in early February and transplanted to cell packs (32 cells per flat) often set bud within 11 weeks and have flowered for us in 10 if the outdoor temperatures around the coldframes has consistently been in the upper 50s or 60s at night; otherwise the plants will

flower in 12-14 weeks at 50-55 night temperatures.

In the garden, plants flower all summer but often appear tired by August.

This is often accompanied by yellow lower foliage, and the plants may die. These notes are based on both the white and rose-pink varieties, though growers to the north and south of northern Illinois have had better success at over wintering these plants than we have. However, the plants

MATRICARIA
Chrysanthemum parthenium
See Chrysanthemum.

Class	Number of Seeds	Seed for 1,000 plants (oz.)	Germination Temperature		Lighting	Days To Germinate	Days Sowing To Transplant	Growing On Temperature	
MYOSOTIS **M. sylvatica** **(Forget-Me-Not)**	45,000/oz. 1,590/g.	1/32	68°–72°	20°–22°	L	8–14	15–25	55°	13°

Ball Culture: Sown in February or March and sold green in the pack in May, forget-me-nots flower sporadically the same season. They are more free-flowering the second year.

We trialed this crop in the same house as our pansies, growing at 48° to 52° F (9° to 11° C) nights. Sowings made in early October, transplanted in packs in late October and into 4-in. (10-cm.) pots in January, showed the first signs of color in March. Plants reached full blooms by mid-April.

Forget-me-nots have proven to be rather short-lived at the Chicago latitude, due primarily to botrytis.

Class	Number of Seeds	Seed for 1,000 plants (oz.)	Germination Temperature		Lighting	Days To Germinate	Days Sowing To Transplant	Growing On Temperature	
OENOTHERA **O. missourensis** **(Ozark Sundrop, Evening Primrose)**	6,500/oz. 227/g.	1/4	70°–80°	21°–26°	L	8–15	18–27	55°	13°
PHYSALIS **P. alkekengi**	17,000/oz. 595/g.	1/8	60°–70°	15°–21°	L/C	7–14	15–25	55°	13°

Ball Culture: Physalis is an annual that comes back each year through reseeding; however, it can become invasive by the second season. Though they do produce flowers, these plants are most valued for their ornamental seed pods which turn a vibrant orange in late summer.

For green pack sales in early May, sow in mid-February. Plants will flower in early July and form pods by early August.

Crop Time (Weeks) Green Packs	Number of Plants		For Same Year Flowering	Blooming Months	Hardiness Zones	Garden Height	Garden Spacing	Staking	Location
	QT./.95 L.	GAL./3.8 L.							
10–12	1	2–3	2, 1C	June–Aug.	3–8	18–20 in. 46–51 cm.	10 in. 25.5 cm.	No	F. Sun
10–12	1	2–3	2, 1C	June–Aug.	3–8	12–15 in. 30.5–38 cm.	8–10 in. 20–25.5 cm.	No	F. Sun
8–9	1	1	2	June–Aug.	3–9	15–20 in. 38–51 cm.	12–15 in. 30.5–38 cm.	No	F. Sun

look very good during the season; they just haven't made it
back as well the second year.

Crop Time (Weeks) Green Packs	Number of Plants		For Same Year Flowering	Blooming Months	Hardiness Zones	Garden Height	Garden Spacing	Staking	Location
10–12	1	1–2	1A, 1C	May–June	5–8	6–8 in. 15–20 cm.	8–10 in. 20–25.5 cm.	No	P. Shade
10–12	1	1	1A, 1C	June–Aug.	4–8	12 in. 30.5 cm.	10 in. 25.5 cm.	No	F. Sun
10–12	1	1–2	2	—	4–8	20–24 in. 51–61 cm.	10–12 in. 25.5–30.5 cm.	No	F. Sun

PERENNIALS

Class	Number of Seeds	Seed for 1,000 plants (oz.)	Germination Temperature		Lighting	Days To Germinate	Days Sowing To Transplant	Growing On Temperature	
			Fahrenheit	Celsius				Fahrenheit	Celsius
PHYSOSTEGIA P. virginiana (False Dragonhead)	18,000/oz. 630/g.	1/8	65°–70°	18°–21°	L/C	7–14	21–28	50°	10°
PLATYCODON P. grandiflorus (Balloon Flower)	32,000/oz. 1,120/g.	1/16	60°–70°	15°–21°	L	7–14	19–28	55°–60°	13°–15°
POPPY Papavar nudicaule (Iceland/Arctic Poppy)	100,000/oz. 3,500/g.	1/64	65°–75°	18°–24°	L	7–12	16–25	50°–55°	10°–13°
Papaver orientale (Oriental Poppy)	95,000/oz. 3,325/g.	1/64	65°–75°	18°–24°	L	7–14	16–25	50°–55°	10°–13°
PYRETHRUM Chrysanthemum coccineum* (Painted Daisy) *See Chrysanthemums.*									
RUDBECKIA R. fulgida (Gloriosa Daisy, Black-Eyed Susan)	25,000/oz. 875/g.	1/16	72°	21°	L/C	7–14	28–38	50°	10°
R. hirta	27,000– 80,000/oz. 945–2,800/g.	1/32	70°	21°	L/C	5–10	20–28	50°	10°

Ball Culture: Germination is irregular, with seedlings appearing intermittently over a 2 week period. Dropping . night temperatures by 8° to 10° F (4° to 6° C) will increase the overall germination rate.

Ball Culture: Platycodons dislike both transplanting and cool weather; plants easily die from an excess of either. In the Chicago area, sowings made in early February and sold green in packs in early May have produced flowering plants by July. Platycodons can be over-wintered in pots but require extra time since they are slow to emerge after dormancy.

Ball Culture: The nudicaule types should be treated as annuals in the Midwest; these plants seldom survive the winter. *The crop time listed for Iceland poppies is for flowering 4-in. (10-cm.) pot sales in mid-May, based on a mid to late-January sowing and transplanted to pots in early March, 2 plants per pot.

Ball Culture: The fulgida types are more difficult to germinate and have a 1 month longer crop time than the hirta varieties. To improve germination for the fulgidas, increase germination temperatures to 75° F or 80° for 5 to 6 days. The fulgida types have displayed excellent winter hardiness in the Chicago area.

Since many hirta types reseed themselves, they should be treated as annuals. These varieties flower profusely, weakening the plants as winter approaches. However, they can survive winters at our Chicago latitude, and we have seen them do well further north. Recent research indicates that B-Nine is effective on the hirta types, but this is provided to show current research and is not a recommendation.

Recommended varieties: Compters Gold is an excellent fulgida variety, similar to Goldsturm, also called Golden Storm. The golden yellow flowers reach 3 in. (7.5 cm.) across, though plants seldom flower profusely the same year from a late winter or spring sowing.

Crop Time (Weeks) Green Packs	Number of Plants		For Same Year Flowering	Blooming Months	Hardiness Zones	Garden Height	Garden Spacing	Staking	Location
	QT./.95 L.	GAL./3.8 L.							
10–12	1	1	1A, 1C	July–Aug.	2–8	24–36 in. 61–91.5 cm.	12 in. 30.5 cm.	No	F. Sun to P. Shade

Sowings made in mid-winter and transplanted green to the field in early May will flower reliably in late July or August.

| 10–13 | 1 | 1–3 | 2, 1A | June–July | 3–8 | 18–24 in. 46–61 cm. | 10–12 in. 25.5–30.5 cm. | No | F. Sun |

Recently a pot dwarf variety was introduced into the trade which combines a uniform habit with excellent pot and garden performance while remaining only 12 to 14 in. tall.

Sentimental Blue bears 2 in. blue flowers on plants that can be used as a pot plant on the front step or as a garden perennial that will dependably over-winter.

Sowings made in early February, transplanted to cell packs in early March and grown on at 55° F (13° C) flowered in mid-June. It is possible that additional colors will be introduced in the next several years.

| 15–17* | 2 | — | 2 | June–July | 6–8 | 10–16 in. 25.5–41 cm. | 10 in. 25.5 cm. | No | F. Sun to P. Shade |
| 10–12 | 1–2 | 2 | 1A, 1C | May–June | 3–8 | 16–36 in. 41–91.5 cm. | 12–15 in. 30.5–38 cm. | No | F. Sun to P. Shade |

| 11–15 | 1 | 2–3 | 1A, 1C | July–Sept. | 3–9 | 20–24 in. 51–61 cm. | 12 in. 30.5 cm. | No | F. Sun |
| 8–10 | 1 | 2–3 | 2 | June–frost | 3–9* | 10–36 in. 25.5–91.5 cm. | 10–12 in. 25.5–30.5 cm. | No | F. Sun |

The hirta types have several varieties of special merit. Sown in January, Goldilocks reaches 10 in. (25.5 cm.) in height with golden orange, double to semi-double flowers appearing in May. Marmalade comes into bloom in 4-in. (10-cm.) pots 7 to 10 days later than Goldilocks from a January sowing.

Taller types such as Double Gold, Rustic Mixture and Single Mixture should be sold green in packs, since these varieties flower later and require long days to bloom.

*Short varieties of R. hirta rarely survives more than one season at the Chicago latitude.

PERENNIALS

Class	Number of Seeds	Seed for 1,000 plants (oz.)	Germination Temperature Fahrenheit	Celsius	Lighting	Days To Germinate	Days Sowing To Transplant	Growing On Temperature Fahrenheit	Celsius
SALVIA **S. x superba** **(Perennial Salvia)** *Also see Bedding, Florist and Foliage Plants section.*	25,000/oz. 875/g.	1/16	72°	22°	L/C	4–8	15–22	55°–60°	13°–15°

Ball Culture: Salvias have proven to be dependably hardy in our trials in the Chicago area. Sown in early April, Stratford Blue can be sold green in packs by early June and will be in flower by July.

Class	Number of Seeds	Seed for 1,000 plants (oz.)	Germination Temperature Fahrenheit	Celsius	Lighting	Days To Germinate	Days Sowing To Transplant	Growing On Temperature Fahrenheit	Celsius
SAPONARIA **S. ocymoides** **(Rock Soapwort)**	16,000/oz. 560/g.	1/8	70°	21°	C	8–10	18–25	55°–60°	13°–15°

Ball Culture: Saponarias are trailing plants. Sowings made in March will flower during May of the following year.

Class	Number of Seeds	Seed for 1,000 plants (oz.)	Germination Temperature Fahrenheit	Celsius	Lighting	Days To Germinate	Days Sowing To Transplant	Growing On Temperature Fahrenheit	Celsius
SCABIOSA **S. caucasica** **(Pincushion Flower)** *Also see Cut Flower section.*	2,400/oz. 84/g.	1	65°–70°	18–21	L	10–18	20–29	55°–58°	13°–14°

Ball Culture: Mid-February sowings, sold green in packs in early May, will flower in late July and August.

Recommended varieties: Fama has offered a fuller habit and a better overall show of color than other varieties we've trialed.

Class	Number of Seeds	Seed for 1,000 plants (oz.)	Germination Temperature Fahrenheit	Celsius	Lighting	Days To Germinate	Days Sowing To Transplant	Growing On Temperature Fahrenheit	Celsius
SEDUM **S. acre** **(Golden Carpet)**	400,000/oz. 14,000/g.	1/256	Day: 85° Night: 70°	Day: 29° Night: 21°	L	8–14	20–29	60°	15°
S. spurium cv. Coccineum **(Dragon's Blood)**	400,000/oz. 14,000/g.	1/256	Day: 85° Night: 70°	Day: 29° Night: 21°	L	8–14	20–29	60°	15°

Ball Culture: Both varieties are excellent garden performers. The S. acre varieties have short, light green leaves with bright-yellow flowers, while the spurium coccineums have rose flowers with darker leaves that turn bronze with the cool weather of fall. Though both types can be invasive, the spurium coccineums exhibit this trait most strongly—one plant may spread as much as 2 ft. in diameter. Sown in early March, the spurium coccineum varieties have flowered for us by early August. Sedums will fill in the first season when used as a ground cover or trailing over a rock wall. Plants provide a superb show of color the following year.

Crop Time (Weeks) Green Packs	Number of Plants		For Same Year Flowering	Blooming Months	Hardiness Zones	Garden Height	Garden Spacing	Staking	Location
	QT./.95 L.	GAL./3.8 L.							
9–11	1	2	2, 1A	June–July	4–7	20–24 in. 51–61 cm.	10 in. 25.5 cm.	No	F. Sun
11–13	1–2	2–3	1A, 1C	May–June	4–8	9 in. 23 cm.	10–12 in. 25.5–30.5 cm.	No	F. Sun
10–12	2	2–3	2, 1A, 1C	July	3–8	24–36 in. 61–91.5 cm.	10 in. 25.5 cm.	No	F. Sun
11–14	2–4	3–5	1A, 2	June	3–8	3–4 in. 7.5–10 cm.	10–12 in. 25.5–30.5 cm.	No	F. Sun
11–14	2–4	3–5	1A, 2	July–Aug.	3–8	3–4 in. 7.5–10 cm.	10–12 in. 25.5–30.5 cm.	No	F. Sun

Class	Number of Seeds	Seed for 1,000 plants (oz.)	Germination Temperature		Lighting	Days To Germinate	Days Sowing To Transplant	Growing On Temperature	
			Fahrenheit	Celsius				Fahrenheit	Celsius
STACHYS S. byzantina (Lamb's Ear)	16,000/oz. 560/g.	1/8	70°	21°	L	8–15	18–24	60°	15°

Ball Culture: With its gray, woolly foliage, this perennial is appropriately nicknamed Lamb's Ear. Mainly used for foliage impact the first season, plants flower the following year from seed.

Class	Number of Seeds	Seed for 1,000 plants (oz.)	Germination Temperature		Lighting	Days To Germinate	Days Sowing To Transplant	Growing On Temperature	
			Fahrenheit	Celsius				Fahrenheit	Celsius
STATICE Limonium latifolia (Wideleaf Sea Lavender, Sea Lavender)	28,000/oz. 980/g.	1/16	65°–75°	18°–24°	—	14–21	24–36	55°	13°
L. tatarica* (German Statice) *Also see the Cut Flowers and Bedding, Florist and Foliage Plants sections.*	22,000/oz. 770/g.	1/8	65°–75°	18°–24°	—	14–21	24–36	55°	13°

Ball Culture: The latifolia and tatarica types are true, long-lived perennials. Limonium latifolia flowers the second year from seed, but take up to 3 years to bloom as freely as other statice types, annual or perennial. The tatarica varieties require extra time to finish. Sown in mid-winter for spring sales. They grow short both in the pack and the first year in the garden.

*Also listed as Goniolimon tataricum and Limonium dumosa.

Class	Number of Seeds	Seed for 1,000 plants (oz.)	Germination Temperature		Lighting	Days To Germinate	Days Sowing To Transplant	Growing On Temperature	
			Fahrenheit	Celsius				Fahrenheit	Celsius
THYMUS T. serphyllum (Mother of Thyme) *Also see Herbs section.*	146,000/oz. 5,110/g.	1/64	71°	21°	C	3–6	12–18	58°	14°

Ball Culture: Seed can be sown direct to 3- or 4-in. (7.5 or 10-cm.) pots and sold green in 16 weeks. Or, sow in February for green pack sales in early May; these trailing plants will be in flower by July and August.

This crop has proven to be dependably hardy in the Chicago area.

Class	Number of Seeds	Seed for 1,000 plants (oz.)	Germination Temperature		Lighting	Days To Germinate	Days Sowing To Transplant	Growing On Temperature	
			Fahrenheit	Celsius				Fahrenheit	Celsius
TRITOMA Kniphofia uvaria (Red Hot Poker)	9,000/oz. 315/g.	1/4	65°–75°	18°–24°	L	21–28	30–45	60°	15°

Ball Culture: Though strong performers, tritomas need ample room to grow and develop. We do not recommend them for pack production unless using larger cell-packs, like 18s. Transplant directly to final pots to allow for good root development. Plants will bloom sporadically from seed the second year but won't reach peak flowering until their third season.

Crop Time (Weeks) Green Packs	Number of Plants		For Same Year Flowering	Blooming Months	Hardiness Zones	Garden Height	Garden Spacing	Staking	Location
	QT./.95 L.	GAL./3.8 L.							
11–13	1	1–2	1A, 1C	June	4–7	12 in. 30.5 cm.	10–12 in. 25.5–30.5 cm.	No	F. Sun to P. Shade
10–11	1	1–2	—	July–Aug.	4–9	18 in. 46 cm.	10 in. 25.5 cm.	**	F. Sun to P. Shade
15–18	1	1–2	1A	June–July	4–9	20 in. 51 cm.	10 in. 25.5 cm.	No	F. Sun to P. Shade

**As plants reach maturity in several years, flower heads get top-heavy and bend to the ground; staking keeps them upright for use as cut flowers.

10–12	1	2–3	1A, 1C, 2	June–Aug.	4–8	2–6 in. 5–15 cm.	10 in. 25.5 cm.	No	F. Sun
—	1	1	—	July–Sept.	5–9	28–36 in. 71-91.5 cm.	15 in. 38 cm.	No	F. Sun

Class	Number of Seeds	Seed for 1,000 plants (oz.)	Germination Temperature Fahrenheit	Celsius	Lighting	Days To Germinate	Days Sowing To Transplant	Growing On Temperature Fahrenheit	Celsius
VERBASCUM V. phoeniceum (Purple Mullein)	198,000/oz. 7,000/g.	1/128	70°–72°	21°–22°	L	4–7	14–21	50°	10°

Ball Culture: An unusual plant whose rosetting leaves reach only 3 to 6 in. in length. However, the long spikes of flowers reach 2-3 ft. the first season from seed. The species blooms purple but the seed offered in the trade is a mixture of seed in which V. phoeniceum has been bred to obtain additional colors. Flowers are often 1 in. long in muted colors of red, pure white, and several shades in between.

Sowings made in early February will be large enough to transplant before the end of the month. The rosetting foliage grows quickly along its width and therefore fills a container in a short period of time. In 9 weeks the plants will fill the flat when using 32 cells per flat.

Class	Number of Seeds	Seed for 1,000 plants (oz.)	Germination Temperature Fahrenheit	Celsius	Lighting	Days To Germinate	Days Sowing To Transplant	Growing On Temperature Fahrenheit	Celsius
VERONICA V. incana (Woolly Speedwell)	221,000/oz. 7,735/g.	1/128	65°–75°	18°–24°	L	7–14	16–25	55°	13°
V. repens (Creeping Speedwell)	221,000/oz. 7,735g.	1/128	65°–75°	18°–24°	L	7–14	16–25	55°	13°
V. spicata (Spike Speedwell)	221,000/oz. 7,735/g.	1/128	65°–75°	18°–24°	L	7–14	16–25	55°	13°

Ball Culture: Of the 3 types, the incanas are the most difficult to germinate and grow. Featuring silvery-gray foliage, they are used most often as accents the first year; plants produce blue flowers the second season.

Vigorous but extremely dwarf, the repens are also noted more for their foliage than their small white to lavender-blue flowers. These plants are ideal for rock gardens and borders.

The spicata varieties grow upright, flowering the same season from seed when sown early in February or March. The spicatas perform best as cut flowers and in perennial borders.

Class	Number of Seeds	Seed for 1,000 plants (oz.)	Germination Temperature Fahrenheit	Celsius	Lighting	Days To Germinate	Days Sowing To Transplant	Growing On Temperature Fahrenheit	Celsius
VIOLA V. cornuta (Horned Violet)	24,000/oz. 840/g.	1/6–1/8	65°–70°	18°–21°	C. Lt.	7–14	15–26	50°–55°	10°–13°
V. tricolor (Johnny Jump-Up)	40,000/oz. 1,400/g.	1/16	65°–70°	18°–21°	C. Lt.	7–14	15–26	50°–55°	10°–13°

Ball Culture: Violas are a small-flowered version of pansies. The tricolor species consistently bloom earlier than pansies and can be sold in either packs or pots. Smaller-flowered than the cornuta types, they are considered annuals since they reseed themselves to come back each year.

January sowings of cornutas will flower in packs by early April; tricolors will flower by late March.

Recommended varieties: In the tricolor species, try Helen Mount and Blue Elf (also called King Henry and Prince Henry). These are by far the most popular cultivars in violas. Among the cornuta species, Blue Perfection (mid-blue). Chantreyland (orange/yellow) and Lutea Splendens (yellow) provide excellent flower color for early spring perennial borders. A newcomer to the trade, the Princess series is growing with new selections being added annually. Plants in this series make excellent pot plants that flower freely 13 to 15 weeks after sowing when grown in a 4 1/2 in. pot.

*The crop times listed are for flowering pack sales.

PERENNIALS

Crop Time (Weeks) Green Packs	Number of Plants		For Same Year Flowering	Blooming Months	Hardiness Zones	Garden Height	Garden Spacing	Staking	Location
	QT./.95 L.	GAL./3.8 L.							
9–11	1	1	2	June–July	5–8	3 ft. 91.5 cm.	12–14 in. 30.5–35.5 cm.	No	F. Sun

If you hold onto your plants for several weeks after they are ready to sell, transplant the seedlings of this crop into 4 or 4½-in. pots instead of cell packs and sell them green.

Regardless of the container you are using, plants will flower readily the first season from seed. In our gardens the plants flower by mid to late June from a February sowing when transplanted to the garden by mid May. When temperatures are warmer than usual, plants die back by the middle or end of July. But trimming the plant back to the ground encourages new growth within a week and possible flowers by Labor Day.

Crop Time (Weeks) Green Packs	Number of Plants		For Same Year Flowering	Blooming Months	Hardiness Zones	Garden Height	Garden Spacing	Staking	Location
12–14	1	1–2	1A	June–July	4–8	12 in. 30.5 cm.	10 in. 25.5 cm.	No	F. Sun to P. Shade
10–12	1	2	1A, 1C	April–May	4–8	2 in. 5 cm.	12 in. 30.5 cm.	No	F. Sun to P. Shade
10–13	1	1–2	2, 1A, 1C	June–July	4–8	24–36 in. 61–91.5 cm.	12 in. 30.5 cm.	No	F. Sun to P. Shade
12–13*	2	—	2	May–June	5–9	7–8 in. 18–20 cm.	10 in. 25.5 cm.	No	F. Sun to P. Shade
11–12*	2	—	2	May–June	5–9	7–8 in. 18–20 cm.	10 in. 25.5 cm.	No	F. Sun to P. Shade

Class	Type	Seed for 1,000 plants (oz.)	Number of Seeds	Germination Temperature Fahrenheit	Celsius	Lighting	Days To Germinate	Days Sowing To Transplant	Growing On Temperature Fahrenheit	Celsius
ANISE Pimpinella anisum	A	1/4	6,000/oz. 210/g.	70°	21°	L/C	10–12	14-21*	60°–62°	15°–17°

Ball Culture: Anise seeds are often used to add a licorice flavor in cooking and baking. Once the plant is established, fresh leaves can be picked and used in salads.

Short lived in the garden, anise planted in late May often flowers, sets seed and then dies by the end of July. High heat and humidity accentuates this process.

*This herb can be sown direct to the final container, 3 to 6 seeds per cell pack, or up to 8 per 4-in. (10-cm.) pot. Thin as needed, transplanting small groups of the extra seedlings to containers.

Class	Type	Seed for 1,000 plants (oz.)	Number of Seeds	Germination Temperature Fahrenheit	Celsius	Lighting	Days To Germinate	Days Sowing To Transplant	Growing On Temperature Fahrenheit	Celsius
BASIL, SWEET Ocimum basilicum	A	1/8	16,000/oz. 560/g.	70°	21°	L/C	5–8	15–18	62°–65°	17°–18°

Ball Culture: The culture information provided here is for green-leaved basil only. Dark or purple-leaved varieties, such as Dark Opal or Purple Ruffles, grow more slowly than the green-leaved types, adding up to 2 more weeks to the crop time. However, green-leaf forms of basil grow quickly from the beginning.

When selling 5- to 6-week-old plants from seed, you can pinch back to the first or second set of leaves upon planting to the garden to encourage branching. The first harvest will be 4 to 5 weeks after planting. Harvests continue at 4-week intervals throughout the summer until the weather turns cool. However,

if planted in early spring (May), plants readily basal branch and do not require a pinch.

Class	Type	Seed for 1,000 plants (oz.)	Number of Seeds	Germination Temperature Fahrenheit	Celsius	Lighting	Days To Germinate	Days Sowing To Transplant	Growing On Temperature Fahrenheit	Celsius
BORAGE Borago officinalis	A	1	2,000/oz 70/g.	70°	21°	C	5–8	14–21	62°–65°	17°–18°

Ball Culture: Borage is an easy-to-grow and quick-to-develop garden plant with a very light aroma. Fresh leaves are used in salads and drinks.

This crop can be sown direct to the final container and sold in packs or small pots. In pots, borage flowers in 8 weeks from a spring sowing.

The plants reseed themselves, but can be short lived.

Class	Type	Seed for 1,000 plants (oz.)	Number of Seeds	Germination Temperature Fahrenheit	Celsius	Lighting	Days To Germinate	Days Sowing To Transplant	Growing On Temperature Fahrenheit	Celsius
CATNIP Nepeta cataria	Pe	1/32	48,000/oz. 1,680/g.	70°	21°	C	5–8	12–16	50°–55°	10°–13°

Ball Culture: As the name implies, cats love to get their paws on this plant! Catnip can also be used in teas and for seasoning. Best sown in the greenhouse and transplanted, catnip plants stand about 8 to 9 in. (20 to 23 cm.) tall when sold at the 7-week stage.

CHIVES
Allium schoenoprasum
See Vegetable section.

HERBS

| Crop Time (Weeks) | | Number Plants Per Pot | Days To Maturity | Garden Spacing | | Garden Height |
Cell Packs	Pots 4-in./10-cm.	4-in./10-cm.		In Rows	Between Rows	
5–6	9–10	2–3	30–45	8–10 in. 20–25.5 cm.	18–24 in. 46–61 cm.	18–24 in. 46–61 cm.
5–6	8–9	1	30–40	12 in. 30.5 cm.	15–18 in. 38–46 cm.	12–24 in. 30.5–61 cm.
5–6	8	1–2	38–50	12–15 in. 30.5–38 cm.	18-24 in. 46-61 cm.	18–24 in. 46–61 cm.
6–7	8–9	1–2	30–45	15–18 in. 38–46 cm.	24 in. 61 cm.	3–3.5 ft. 91.5–107 cm.

Class	Type	Seed for 1,000 plants (oz.)	Number of Seeds	Germination Temperature		Lighting	Days To Germinate	Days Sowing To Transplant	Growing On Temperature	
				Fahrenheit	Celsius				Fahrenheit	Celsius
CORIANDER Coriandrum sativum	A	1/2	3,000/oz. 105/g.	60°	15°	L/C	7–10	14–18	55°–58°	13°–14°
DILL Anethum graveolens	A	1/8	27,000/oz. 945/g.	60°	15°	L	5–8	10–15	55°–58°	13°–14°
FENNEL, SWEET Foeniculum vulgare	A	1/4	8,000/oz. 280/g.	70°	21°	C	7–10	12–15	55°–60°	13°–15°
HOREHOUND Marubium vulgare	Pe	1/16 to 1/8	24,000/oz. 840/g.	70°	21°	C	12	25-28	55°–58°	13°–14°
LAVENDER, ENGLISH Lavandula angustifolia	Pe	1/8	25,000/oz. 875/g.	65°–75°	18°–24°	L	14–21	20–32	58°–60°	14°–15°

Ball Culture: Also called Chinese parsley or cilantro, coriander's leaves are used in cooking, while the dried seeds add zest to pickles, salads and Mexican salsa.

Grow in 4-in. (10-cm.) pots or large cell packs and sell green.

Following the culture above, plants will be 6 to 8 in. (15 to 20 cm.) tall when sold in packs. Plants will flower 10 to 12 weeks after sowing.

Ball Culture: Dill is an upright annual used primarily for pickling. Occasionally, it is also sold as a fresh cut flowers at roadside markets.

Transplant 1 or 2 seedlings to each cell pack. After 5 to 6 weeks, the plants will be 6 to 7 in. (15 to 18 cm.) tall and quickly become overgrown. If not potted up once established in the cell pack, plants will be in flower within 10 weeks.

Recommended varieties: Dill sold in the trade is a tall, robust crop growing to 3 or 4 ft. tall. This is the most common selection. Another variety, Fern Leaf, has fern-like foliage on plants to 2 ft. or less. Fern Leaf is an All America Award winner.

Ball Culture: Leaves of sweet fennel closely resemble those of dill, and the culture for the 2 herbs is very similar. Like anise, fennel has a licorice flavor; the seeds and leaves are used for pickles, Italian breads, salads and more.

Sow direct to the final container or transplant individual seedlings. If using cell packs, transplant one seedling per cell, not allowing it to become root bound. A well-rooted, or established, crop in cell packs usually occurs in approxi-

mately 7 weeks. It becomes root-bound about 10 days later. When fennel is grown in 4-in. pots, allowing it to become root-bound is not as critical.

Ball Culture: Featuring a strong, minty taste, horehound is a favorite flavoring for candies, cough syrups and honey. This crop is somewhat slower to grow and develop than many other herbs.

Horehound will be 5 to 7 in. (13 to 18 cm.) when sold in packs, following the culture above.

Ball Culture: An aromatic, sweetly scented herb, lavender is a favorite ingredient in potpourri and sachets.

Proven through years of trialing herbs in our West Chicago location, lavender and rosemary are by far the latest to germinate and develop when compared to any other crop listed in this guide. Though both crops can be transplanted in 28 days, you may prefer to add another week in the sowing tray for larger plants. Seedlings grow upright and will not branch in cell packs, so give them plenty of time to develop.

Plants will be 4 to 5 in. (10 to 13 cm.) tall when sold at the 13- to 14-week stage.

A new variety, Lavender Lady, is an All America Award winner, growing to 4 or 5 in. tall. It has better seed quality than the standard variety with 15% to 25% better germination. Plants also will grow and finish faster than the standard variety.

| Crop Time (Weeks) | | Number Plants Per Pot | Days To Maturity | Garden Spacing | | Garden Height |
Cell Packs	Pots 4-in./10-cm.	4-in./10-cm.		In Rows	Between Rows	
6–7	8–9	1–2	38–48	15–18 in. 38–46 cm.	18–24 in. 46–61 cm.	20-24 in. 51–61 cm.

Coriander is short lived in the garden, and often dies out by August when planted to the garden from a cell pack in late May.

6–7	8–9	1–2	30–45	12–15 in. 30.5–38 cm.	24 in. 61 cm.	24–36 in. 61–91.5 cm.
6–7	8–9	1–2	50–65	15–18 in. 38–46 cm.	24 in. 61 cm.	4–5 ft. 1.2–1.5 m.

Once flowering begins, fennel loses its sweet flavor and becomes bitter. Following the culture above, the plants will be 6 to 8 in. (15 to 20 cm.) tall when sold in the pack.

10–11	13–14	2	30–40	12–15 in. 30.5–38 cm.	15–18 in. 38-46 cm.	20-28 in. 51–71 cm.
15–18	18–20	1–2	30–60	15–18 in. 38–46 cm.	15–18 in. 38–46 cm.	12–26 in. 30.5–66 cm.

HERBS

Class	Type	Seed for 1,000 plants (oz.)	Number of Seeds	Germination Temperature		Lighting	Days To Germinate	Days Sowing To Transplant	Growing On Temperature	
				Fahrenheit	Celsius				Fahrenheit	Celsius
MARJORAM, SWEET Origanum majorana	A	1/28	160,000/oz. 5,600 g.	70°	21°	C. Lt.	4–8	10-14	55°–58°	13°–14°

Ball Culture: A popular seasoning for main dishes, sauces and soups, marjoram also scents sachets and soaps. Both fresh and dried leaves are used. This trailing herb fills in well and basally branches in either cell packs or 4-in. (10-cm.) pots. Using the culture above, marjoram plants will be about 4 in. (10 cm.) tall when sold.

| **OREGANO** Origanum vulgare | Pe | 1/128 | 245,000/oz. 8,575/g. | 70° | 21° | C | 4–8 | 16–18 | 50°–55° | 10°–13° |

Ball Culture: Found in many Italian recipes, oregano is another trailing herb with a strong aroma. Grown in cell packs u sing the culture above, plants will be about 4 in. (10 cm.) tall when sold.

| **PARSLEY** Petroselenium crispum *See Vegetable section.* | | | | | | | | | | |

| **PEPPERMINT** Mentha x piperita | Pe | 1/256 | 472,000/oz. 16,520/g. | 70°–75° | 21°–24° | C. Lt. | 12 | 25–29 | 55°–58° | 13°–14° |

Ball Culture: The leaves of peppermint and its relative, spearmint, are used in aromatics, and the oils flavor condiments, candies and medicines. Both are trailing plants that grow to about 5 in. (13 cm.) tall when sold in packs, using the culture above.

*Fast-developing crops, peppermint and spearmint trail and re-root frequently, creating new plants along the way. Both can be harvested for the first time shortly after planting to the garden.

Note: Plants are weedy and will take over a small patch quickly and kill out any small plants as they spread.

| **ROSEMARY** Rosmarinus officinalis | Pe | 1/4 | 21,000/oz. 735/g. | 70° | 21° | L | 10-15 | 21–28 | 55°–58° | 13°–14° |

Ball Culture: A pine-scented herb, rosemary is used as an aromatic and as a flavoring in soups, meats and other foods, especially Italian dishes.

Because this crop tends to have low germination, freeze unused seed between sowings to aid in germination. Alternating day and night temperatures of 70°/55° F (21°/13° C) may also increase germination. Rosemary is slow to develop and grow, so be sure plants are of good size before transplanting.

| **SAGE** Salvia officinalis | Pe | 1/2 | 3,000/oz. 105/g. | 70° | 21° | C | 6–10 | 20-27 | 55°–58° | 13°–14° |

Ball Culture: A common ingredient in sausage, stuffing and sauce recipes, sage is particularly popular in Italian cooking. Following the culture above, plants will be about 4 to 5 in. (10 to 13 cm.) tall in the pack when sold.

HERBS

Crop Time (Weeks)		Number Plants Per Pot	Days To Maturity	Garden Spacing		Garden Height
Cell Packs	Pots 4-in./10-cm.	4-in./10-cm.		In Rows	Between Rows	
8–9	10–11	1–2	30–40	12–15 in. 30.5–38 cm.	12–18 in. 30.5–46 cm.	12–24 in. 30.5–61 cm.
10–11	12–13	1–2	35–45	10–12 in. 25.5–30.5 cm.	15–20 in. 38–51 cm.	15–24 in. 38–61 cm.
9–10	11–12	1-2	*	15–18 in. 38–46 cm.	24 in. 61 cm.	24–28 in. 61–71 cm.
12–13	15	1–2	60–70	10–12 in. 25.5–30.5 cm.	12–15 in. 30.5-38 cm.	10–15 in. 25.5–38 cm.
9–10	11–12	1	38–50	12–15 in. 30.5–38 cm.	20–24 in. 51-61 cm.	15-20 in. 38–51 cm.

HERBS

Class	Type	Seed for 1,000 plants (oz.)	Number of Seeds	Germination Temperature		Lighting	Days To Germinate	Days Sowing To Transplant	Growing On Temperature	
				Fahrenheit	Celsius				Fahrenheit	Celsius
SAVORY, SUMMER Satureia hortensis	A	1/32–1/16	52,000/oz. 1,820/g.	70°	21°	L	12–15	16–20	60°–62°	15°–17°

Ball Culture: Used fresh or dried, the leaves of savory have a spicy, peppery flavor ideal for salads and dressings. Savory is a fast-growing plant that can quickly get out of hand; unless you're using larger cell packs, this crop may be better grown in 3-or 4-in. (7.5 or 10-cm.) pots. The stems will be pure white and the growth may be somewhat stringy.

Class	Type	Seed for 1,000 plants (oz.)	Number of Seeds	Germination Temperature		Lighting	Days To Germinate	Days Sowing To Transplant	Growing On Temperature	
SPEARMINT Mentha spicata	Pe	1/256	472,000/oz. 16,520/g.	70°–75°	21°–24°	C. Lt.	12	25–29	55°–58°	13°–14°

Ball Culture: See the culture for peppermint.

* Fast-developing crops, spearmint and peppermint trail and re-root frequently, creating new plants along the way. Both can be harvested for the first time shortly after planting to the garden.

Class	Type	Seed for 1,000 plants (oz.)	Number of Seeds	Germination Temperature		Lighting	Days To Germinate	Days Sowing To Transplant	Growing On Temperature	
THYME, CULINARY Thymus vulgaris *Also see Perennial Plants section.*	Pe	1/64	124,000/oz. 4,340/g.	70°	21°	C	3–6	12–18	55°–58°	13°–14°

Ball Culture: Thyme is one of the premiere herbs in Italian sauces and French recipes. Following the culture above, plants will be about 4 in. (10 cm.) tall when sold in packs.

| Crop Time (Weeks) | | Number Plants Per Pot | Days To Maturity | Garden Spacing | | Garden Height |
Cell Packs	Pots 4-in./10-cm.	4-in./10-cm.		In Rows	Between Rows	
7–8	9–10	1–2	30–45	15–18 in. 38–46 cm.	18–24 in. 46–61 cm.	15-20 in. 38–51 cm.
9–10	11–12	1-2	*	15–18 in. 38–46 cm.	24 in. 61 cm.	24–28 in. 61–71 cm.
11–12	13–15	1–2	40–55	10–12 in. 25.5–30.5 cm.	18–24 in. 46-61 cm.	10–12 in. 25.5–30.5 cm.

HERBS

Class	Seed for 1,000 plants (oz.)	Number of Seeds	Germination Temperature		Lighting	Days To Germinate	Days Sowing To Transplant	Growing On Temperature	
			Fahrenheit	Celsius				Fahrenheit	Celsius
ARTICHOKE Cynara scolymus	3	600/oz. 21/g.	70°–75°	21°–24°	L/C	18–21	28–35	62°–65°	17°–18°

Ball Culture: Soak seed overnight. Artichokes are most commonly grown from suckers and do best in areas that are relatively frost-free with cool, moist summers. Artichokes have been grown in a number of states, but California has the best conditions overall. Temperatures of 28° to 30° F (-2° to -1° C) will freeze out the flowers, while temperatures below 28° F (-2° C) will kill the plants.

If flower buds are not picked green, they develop into thistle-like blooms in colors ranging from lavender to purple, 7 in. (18 cm.) across. These are sometimes used as fresh cut flowers.

BEET Beta vulgaris	1-1.25	1,200/oz. 42/g.	72°	22°	C. Lt.	15	—	55°–60°	13°–15°

Ball Culture: Sow seed direct to the final container to avoid transplanting from a seed flat. In the northern U.S., make sowings 3 to 4 weeks apart for sales from early April to mid-June.

In the southern U.S.: Sow seed 3 to 4 weeks apart for green pack sales from early September to March.

The seed sown is actually dried fruit with up to 6 seeds inside each one. Therefore, sow only 1 to 2 seeds per cell. As with any root crops, the best taste and overall performance is from crops grown under cool weather.

BROCCOLI Brassica oleracea var. italica	1/4	6,000–9,000/oz. 210–315/g.	70°	21°	C	10	10–21	50°–60°	10°–15°

Ball Culture: Broccoli requires moist, cool conditions. For further cultural information, see the comments for cabbage.

Recommended varieties: All Ball Seed varieties are F_1 hybrids. Premium Crop (65 days) and Packman (56 days) are ideal for home gardening and freezing, with excellent side shoot development and large, 8- to 9-in. (20- to 23-cm.) heads.

BRUSSELS SPROUTS Brassica oleracea var. gemmifera	1/4	7,000/oz. 245/g.	70°	21°	C	10	10-14	50°–60°	10°–15°

Ball Culture: See the comments for cabbage.

CABBAGE Brassica oleracea var. capitata	1/4	7,000/oz. 245/g.	70°	21°	C	10	10-14	50°–60°	10°–15°

Ball Culture: Cabbage and its cousins are collectively called Cole Crops. Besides cabbage, this group includes broccoli, brussels sprouts, cauliflower, collard, kale and kohlrabi. Germination occurs quickly, usually within 4 to 5 days, and all germination occurs within 7 to 10 days of sowing. Upon germination, reduce night temperatures to 60° to 62° F (16° to 17° C), and place under lights, if available. Additional lighting and lower temperatures help pre-vent stretching and tone up plants before transplanting. If plants have stretched before transplanting, bury them up to their seed leaves when transferring them to the final container. Though it's not usually necessary, apply a preventative fungicide in growing areas where fungus diseases are prevalent. After 5 to 7 days, drop the night temperatures gradually to 50° to 55° F (10° to 13° C) to harden plants for sale.

In the northern U.S., have plants ready for sale from early April to mid or late May. Sow again in June for green pack sales in July for fall harvest.

In the southern U.S.: Sales should be planned for fall and winter only. Have plants ready for sale from September to February, except kohlrabi, cauliflower and collard. Kohlrabi and collard should be sold from late February to early April,

Crop Time (Weeks)		Number Plants Per Pot	Days To Maturity	Spacing		Gardening	
Cell Packs	Pots 4-in./10-cm.	4-in./10-cm.		In Rows	Between Rows	Height	Width/Spread
—	12–16	1	*	3–4 ft. 91.5–122 cm.	6 ft. 1.8 m.	3–4 ft. 91.5–122 cm.	5–6 ft. 1.5–1.8 m.

*Crop time from transplanting or sowing direct to the field is approximately 1 year, unless seed is started under glass and the season is both long and warm.

| 3–4 | — | — | 50–65 | 2–4 in. 5–10 cm. | 18–30 in. 46–76 cm. | 12–14 in. 30.5–35.5 cm. | 12–14 in. 30.5-35.5 cm. |

Beets are biennials which can be sold as green plants in the late summer for overwintering in home gardens in the northern and southern U.S. Plants should be in the garden no later than mid-September at Chicago or Boston latitudes, earlier in areas farther north. Southern home gardeners should grow this crop over the winter for harvest in the spring.

| 6–8 | 8–9 | 1 | 50–58 | 2–3 ft. 61–91.5 cm. | 3 ft. 91.5 cm. | 2–2.5 ft. 61–76 cm. | 2 ft. 61 cm. |

| 6–8 | 8–9 | 1 | 80–90 | 2–3 ft. 61–91.5 cm. | 3 ft. 91.5 cm. | 3–4.5 ft. 91.5–137 cm. | 2–2.5 ft. 61–76 cm. |

| 6–8 | 8–9 | 1 | 40-90* | 3 ft. 91.5 cm. | 3 ft. 91.5 cm. | 12–14 in. 30.5–35.5 cm. | 2–3 ft. 61–91.5 cm. |

and again from September to November. Cauliflower plants should be sold January to early March, and again August through mid-October.

*This does not include Chinese cabbage.

Recommended varieties: F₁ crops–Emerald Cross (67 days) produces heads as large as 6 lbs. (2.7 kg.); Golden Cross (40 days) has blue-green heads up to 2¹/₂ lbs. (1.1 kg.). Non-hybrid crops–Early Flat Dutch (80 days) has heads up to 4 lbs. (1. 8 kg.).

VEGETABLES

Class	Seed for 1,000 plants (oz.)	Number of Seeds	Germination Temperature		Lighting	Days To Germinate	Days Sowing To Transplant	Growing On Temperature	
			Fahrenheit	Celsius				Fahrenheit	Celsius
CAULIFLOWER Brassica oleracea var. botrytis	1/4	7,000/oz. 245/g.	70°	21°	C	10	10–14	50°–60°	10°–15°
CELERY Apium graveolens var. dulce	1/32	71,000/oz. 2,485/g.	70°	21°	C	10-21	15–28	60°	15°
CHIVES Allium schoenoprasum	1/16	33,000/oz. 1,155/g.	70°	21°	C	14	—	60°	15°
COLLARD Brassica oleracea var. acephala	1/4	7,000/oz. 245/g.	70°	21°	C	10	10–14	50°-60°	10°–15°
CORN, SWEET Zea mays var. saccharata	8	150/oz. 5/g.	70°–75°	21°-24°	C	5	—	60°–62°	15°–17°
CUCUMBER Cucumis sativus	1.5	1,000/oz. 35/g.	72°	22°	C	7	—	60°-62°	15°–17°

Ball Culture: Plants will 'button up' instead of producing large, firm heads of white florets if stressed during growth by limited or excessive moisture, too hot or cool temperatures, or other adverse weather conditions. For further cultural information, see the comments for cabbage.

Recommended varieties: Snow Crown (50 days) is our most popular F_1 variety, with pure white heads up to 2 lbs. (.9 k g.) in weight. Non-hybrid Early Snowball A (60 days) has medium-sized heads with smooth white florets.

Ball Culture: These cool-tolerant plants are somewhat difficult to grow. Celery does best in cool conditions with moist to wet soil. The plants need plenty of time—up to 140 days for late-season varieties—so have transplants available in early May in northern greenhouses, and from January to March in the southern U.S.

Ball Culture: Sow seed direct to the final container. Since chives grow as singular strands that do not branch, sow 8 to 10 seeds per cell pack, 10 to 15 to a 4-in. (10-cm.) pot. If plants get too 'weedy' in appearance, shear them off 1 to 2 in. (2.5 to 5 cm.) above the soil and they will produce new shoots.

Ball Culture: See the comments for cabbage.

Ball Culture: Sow seed direct to the final container, and sell within 3 to 4 weeks.

Ball Culture: Sow seed direct to the final container, using either a Jiffy-pot or peat-based container so the container and roots can be buried without disturbing the root system.

Sow 2 seeds per pot if using the 3-in. (7.5 cm.) Jiffy-pot, and do not thin. When selling these, tell customers to bury the entire container up to the leaves of the plants.

Recommended varieties: F_1 hybrid pickling types—both Bush Pickle (48 days) and Lucky Strike (52 days) have 4- to 5-in. (10- to 13-cm.) long fruit.

| Crop Time (Weeks) | | Number Plants Per Pot | Days To Maturity | Spacing | | Gardening | |
Cell Packs	Pots 4-in./10-cm.	4-in./10-cm.		In Rows	Between Rows	Height	Width/Spread
6–8	8–9	1	45-68	3 ft. 91.5 cm.	3 ft. 91.5 cm.	15–18 in. 38–46 cm.	2–3 ft. 61-91.5 cm.
8–10	—	—	80-140	6–10 in. 15–25.5 cm.	2–3 ft. 61–91.5 cm.	2–3 ft. 61–91.5 cm.	8–14 in. 20–35.5 cm.
—	—	—	80	—	—	12–14 in. 30.5–35.5 cm.	4–6 in. 10-15 cm.
6–8	8–9	1	70	3 ft. 91.5 cm.	3 ft. 91.5 cm.	1–2 ft. 30.5–61 cm.	18–24 in. 46–61 cm.
1–2	—	—	75–90	12–18 in. 30.5–46 cm.	2–3 ft. 61–91.5 cm.	5–6 ft. 1.5–1.8 m.	2–3 ft. 61–91.5 cm.
—	3–5 (Jiffy)	2–3	48–70	2–3 ft. 61–91.5 cm.	4–6 ft. 1.2–1.8 m.	12–15 in. 30.5-38 cm.	6 ft. 1.8 m.

Bush Pickle also makes an excellent container plant. F_1 hybrid slicers—Fanfare (63 days) and Salad Bush (57 days) produce fruit up to 7 in. (18 cm.) long.

VEGETABLES

VEGETABLES

Class	Seed for 1,000 plants (oz.)	Number of Seeds	Germination Temperature Fahrenheit	Germination Temperature Celsius	Lighting	Days To Germinate	Days Sowing To Transplant	Growing On Temperature Fahrenheit	Growing On Temperature Celsius
EGGPLANT Solanum melongea var. esculentum	1/4	6,000/oz. 210/g.	70°–75°	21°–24°	L/C	7–14	12–18	60°–62°	15°–17°

Ball Culture: A member of the same family as tomatoes and peppers, eggplant does not appreciate cool weather. In the northern U.S., sales can begin in mid-April and continue through the first weeks in June. For early sales in areas where frost is still a danger, we recommend covering the plants at night.

In the southern U.S.: Sales can begin in late February and continue until early July.

Recommended varieties: F₁ hybrids—Ichiban Imp. (54 days) is one of our most popular varieties, with long and slender, oriental-type fruit. Another favorite, Satin Beauty (65 days) is an early-bearing type with more rounded, deep-purple fruit.

Class	Seed for 1,000 plants (oz.)	Number of Seeds	Germination Temperature Fahrenheit	Germination Temperature Celsius	Lighting	Days To Germinate	Days Sowing To Transplant	Growing On Temperature Fahrenheit	Growing On Temperature Celsius
KALE Brassica oleracea (acephala)	1/4	6,000/oz. 210/g.	70°	21°	C	10	10-14	50°–60°	10°–15°

Ball Culture: See the comments for cabbage.

| **KOHLRABI** Brassica oleracea var. gongylodes | 1/4 | 6,000/oz. 210/g. | 70° | 21° | C | 10 | 10-14 | 50°–60° | 10°–15° |

Ball Culture: See the comments for cabbage.

| **LETTUCE** Lactuca sativa | 1/8 | 15,000/oz. 525/g. | 70° | 21° | L/C | 7 | * | 55°–58° | 13°-14° |

Ball Culture: In the northern U.S., have plants ready for spring sales in early April and again in August for loose-leaf types. Lettuce plants can withstand temperatures to just above freezing if properly hardened off.

In the southern U.S.: Planting can begin in late January to mid-March, and again in September once the hottest part of the summer has passed. To determine when sowing should begin, count the number of weeks for crop time backward from the start of sales.

*Sow lettuce seed direct to the final container, such as a Jiffy-pot, standard pot or cell pack. Use several seeds per container, and thin as necessary. Lettuce can also be sown in an open flat and transplanted to the final container once the seedlings are approximately 1 in. (2.5 cm.) tall.

| **MUSKMELON** Cucumis melo var. reticulatis | 2 | 700–900/oz. 24–31/g. | 75°–80° | 24°–26° | C | 10 | * | 65°–70° | 18°–21° |

Ball Culture: Muskmelons are warm-weather plants and hate cold conditions. In the northern U.S., have plants ready for sale once the danger of frost has passed, through early to mid-June. In extended warm-weather regions in the south, have plants available for sale in March and April.

Recommended varieties: Since muskmelons are long season plants that won't tolerate frost, choose only varieties that fit your growing region. Excellent choices for home garden use include Ambrosia (88 days), a heavily netted variety with fruit up to 4 lbs. (1.8 kg.); Ball 1776 (84 days), our sweetest and best-tasting muskmelon that reaches up to 5 lbs. (2.3 kg.); and Pulsar (84 days), a high-quality variety that produces melons up to 5.5 lbs. (2.5 kg.) with heavy netting and deep ribs.

*Sow seed direct to the final container. Like other vine crops, muskmelon should be sown in a Jiffy-pot, 2 to 3 seeds per pot. Upon germination, leave all 3 plants in the container, or thin as needed. Do not attempt to sow to an open flat and transplant. Numerous transplantings will damage not only the roots but also the eventual yields. Transplant Jiffy-pot and all into the garden.

| Crop Time (Weeks) | | Number Plants Per Pot | Days To Maturity | Spacing | | Gardening | |
Cell Packs	Pots 4-in./10-cm.	4-in./10-cm.		In Rows	Between Rows	Height	Width/Spread
6–8	8–9	1	54–80*	2-3 ft. 61–91.5 cm.	3–4 ft. 91.5-122 cm.	1.5–2 ft. 46–61 cm.	14–18 in. 35.5–46 cm.

*These crop times reflect all the various types and shapes of eggplants. Check the Ball Seed Catalog for specific crop times of particular varieties.

6–8	8–9	1	45–55	2–3 ft. 61–91.5 cm.	3 ft. 91.5 cm.	18–24 in. 46–61 cm.	2-2.5 ft. 61–76 cm.
6–8	8–9	1	45-55	10–12 in. 25.5–30.5 cm.	2–3 ft. 61-91.5 cm.	15–18 in. 38–46 cm.	18–24 in. 46–61 cm.
4–5	6–9	1–2	**	10–12 in. 25.5–30.5 cm.	18–24 in. 46-61 cm.	**	**

**Maturity times and heights differ by type: Loose-leaf varieties take 40 to 50 days, reaching up to 15 in. (38 cm.) tall and 12 to 14 in. (30.5 to 35.5 cm.) across. Bibb or Buttercrunch types take 70 to 85 days and stand up to 12 in. (30.5 cm.) tall and 12 to 14 in. (30.5 to 35.5 cm.) across. Head lettuce takes up to 100 days to develop and ranges from 12 to 15 in. (30.5 to 3 8 cm.) tall and up to 18 in. (46 cm.) across.

—	—	—	72–90** (Jiffy)	1–2 ft. 30.5–61 cm.	5–6 ft. 1.5–1.8 m.	10–14 in. 25.5–35.5 cm.	4–6 ft. 1.2–1.8 m.

**Maturity times are for netted or ribbed melons, and don't include Honeydew varieties which can take 90 to 110 days.

Class	Seed for 1,000 plants (oz.)	Number of Seeds	Germination Temperature		Lighting	Days To Germinate	Days Sowing To Transplant	Growing On Temperature	
			Fahrenheit	Celsius				Fahrenheit	Celsius
OKRA Abelmoschus esculentus	3	500/oz. 17/g.	70°	21°	C. Lt.	7-14	14–18	60°–62°	15°–17°

Ball Culture: Okra is related to hibiscus and requires a warm season to produce. In fact, it will not tolerate weather below 54° F (12° C), so keep plants warm. Soak seed overnight but for no more than 24 hours prior to sowing into a deep seed tray.

In the northern U.S., have plants ready for sale once the temperatures reach 60° F (16° C) and nights are above 55° F (13° C). This crop is not recommended for the very

| **ONION** Allium cepa | 1/4 | 9,000/oz. 315/g. | 70° | 21° | L/C | 10 | — | 55°-60° | 13°–15° |

Ball Culture: Sow seed direct to the final container using cell packs, pots or open flats. Thin seedlings to leave 1 to 2 per cell, or 1 in. (2.5 cm.) between plants in an open flat.

Take the pulled seedlings and plant them in cell packs or other containers.

In the northern U.S., have the bulbing or bunching type of onion plants ready for sale from April to June.

| **PARSLEY** Petroselenium crispum | 1/8 | 18,000/oz. 630/g. | 70° | 21° | C | 15–20 | 21–28 | 60°–62° | 15°–17° |

Ball Culture: This crop can be sown direct to the final container or transplanted into packs or pots from a seed flat. If sowing direct, use a number of seeds for each container and thin seedlings as needed. If germination for a particular variety is irregular and low, put seed on moistened sand and chill for 1 to 2 days, then sow as normal. Some growers also soak the seed overnight to increase germination. Either or both of these methods may increase germination, though parsley is not particularly difficult to germinate.

In the northern U.S., plants can be sold from the last frost through mid-June.

| **PEPPER** Capsicum annuum | 1/2 | 4,000/oz. 140/g. | 72° | 22° | L/C | 10 | 21–26 | 62° | 17° |

Ball Culture: Peppers are a warm-weather crop which performs best once the soil and night air temperatures reach 58° F (14° C) or higher. In the northern U.S., plants should be ready for sale in mid-April for homeowners who plant into the garden with night protection. Sales will hit their peak in May and continue on until early June.

In the southern U.S.: The best plants come from garden plantings between mid-February and April. While the deep south (such as Florida) can get a fall crop in, most of the lower tier of states will get little yield from a fall planting.

Recommended varieties: The pepper family is quite extensive and diverse, including well over 50 varieties of varying sizes and colors, and ranging from the mildest of blocky-shaped bells to the red-hot taste of Fire!. Among our best selling varieties are: Better Belle (65 days), a blocky green bell with a sweet, mild taste—ideal for cooking, stuffing

| **POTATO** Solanum tuberosum | 1/32 | 50,000/oz. 1,750/g. | 60°–65° | 15°–18° | C. Lt. | 7–14 | 20–28 | 60° | 15° |

Ball Culture: Germination occurs in 2 flushes and will be low if temperatures are kept cooler than 55° F (13° C).

In the northern U.S., plants can be sold after all danger of frost has passed until mid-June. Potatoes from seed are not as large as those produced from tubers, and are best reserved for fresh use since the skin is thin and the potatoes damage easily.

In the southern U.S.: Plant after all danger of frost, from early March to May. Potatoes can be dug in early June and July. For a fall/winter crop in the south, have plants ready for sale from early to mid-September, for harvest in late December to early January.

Information is based on Explorer.

| Crop Time (Weeks) | | Number Plants Per Pot | Days To Maturity | Spacing | | Gardening | |
Cell Packs	Pots 4-in./10-cm.	4-in./10-cm.		In Rows	Between Rows	Height	Width/Spread
4–6	8–9	1	60	1–2 ft. 30.5–61 cm.	3-4 ft. 91.5–122 cm.	3–4 ft. 91.5–122 cm.	2–3 ft. 61–91.5 cm.

northern border of the U.S. or Canada, unless 80 days of 60° F (16° C) or warmer weather are expected.

In the southern U.S.: Okra can be sold anytime from mid-March to July.

| 6–8 | 8–9 | — | 95-130 | 2–4 in. 5–10 cm. | 2-3 ft. 61–91.5 cm. | 18–24 in. 46–61 cm. | 3–4 in. 7.5–10 cm. |

In the southern U.S.: It is best to have onions ready for sale anytime from August to mid-November.

| 9–10 | 10–11 | 3-6 | 70–80 | 6–10 in. 15–25.5 cm. | 18–24 in. 46–61 cm. | 12–15 in. 30.5–38 cm. | 12–14 in. 30.5–35.5 cm. |

In the southern U.S., plants can be sold from February to early April.

Parsley is a biennial so is able to overwinter with some protection in latitudes around Chicago or Boston and further south. Harvest the leaves the first season until frost, and the second season until flowers form; then discard the plants. Once flower buds form, the leaves turn bitter.

If plants are left once sales are over, pot them in 5- or 6-in. (13- to 15-cm.) pots, and sell as indoor window garden plants for fresh snippings.

| 6–8 | 8–9 | 1 | 60-80 | 18–24 in. 46–61 cm. | 3 ft. 91.5 cm. | 24–30 in. 61–76 cm. | 15–20 in. 38–51 cm. |

and snacking; and California Wonder (75 days), a non-hybrid, blocky green bell variety with thick walls and a flavor perfect for salads and snacking. For a superb yellow bell, try Sun Bell (70 days), a bright, golden yellow pepper highlighted by a delicious, sweet taste. For a hot touch, try Mabanero (95 days), a bell variety with a spicy, chili bite.

Another hot pepper, Super Chili (75 days) has tapered fruit up to 3 in. (7.5 cm.) long. See the Ball Seed Catalog for a chart describing all the varieties we sell.

| 6–8 | 8–10 | 1–2 | 90–120 | 10–12 in. 25.5–30.5 cm. | 2–3 ft. 61-91.5 cm. | 12-15 in. 30.5–38 cm. | 12–14 in. 30.5–35.5 cm. |

VEGETABLES

127

Class	Seed for 1,000 plants (oz.)	Number of Seeds	Germination Temperature		Lighting	Days To Germinate	Days Sowing To Transplant	Growing On Temperature	
			Fahrenheit	Celsius				Fahrenheit	Celsius
PUMPKIN Cucurbita maxima	6.5	200/oz. 7/g.	72°	22°	C	7	*	62°	17°
SQUASH Cucurbita maxima	6.5	175–300/oz. 6–10/g.	72°	22°	C	10	—	62°	17°
STRAWBERRIES Fragaria x hybrida	1/32	60,000/oz. 2,100/g.	65°	18°	C. Lt.	21–28	28–38	60°-62°	15°–17°
TOMATO Lycopersicon lycopersicum	1/8	9,000– 11,000/oz. 315–385/g.	70°-75°	21°–24°	C	7–14	12–18	62°–65°	17°–18°
WATERMELON Citrullus lanatus	6.5	300/oz. 10/g.	70°–72°	21°–22°	C	8	*	65°–70°	18°–21°

PUMPKIN

Ball Culture: Sow direct to the final container using Jiffy-pots or comparable containers so pot and all can be planted.

Pumpkins like warm or hot weather, so plants should be ready for sale after all danger of frost has passed. In the north, sales will continue until mid or late May. Remember that pumpkins require a minimum of 90 days to yield harvestable fruit, so customers should plant the pumpkins as soon as the ground is warm enough.

SQUASH

Ball Culture: There are 2 kinds of squash, the bush type and the trailing type. Summer and winter squash may be either type. Like pumpkins, melons and cucumbers, squash resent transplanting. Therefore, sow seed direct to the final container using a Jiffy-pot or comparable container so pot and all can be planted.

Squash reaches the salable stage very quickly. Have plants ready for sale after all danger of frost has passed. In the northern U.S., plants should be sold from mid-April to May.

STRAWBERRIES

Ball Culture: There is only one ever-bearing strawberry in this class called SweetHeart. (Note: we do not recommend growing it in packs.) Once SweetHearts are in flower (9 to 10 weeks after sowing), the tendrils have already formed and will root down in other cell packs, making the plants more difficult to divide. For 4-in. (10-cm.) pots, transplant from seed flats into packs, or transplant direct to the final pots. For hanging basket sales, put 10 to 15 plants in each 10-in. (25.5-cm.) basket 15 to 17 weeks prior to sale.

TOMATO

Ball Culture: Tomatoes are the top-selling vegetable in the U.S. Easily grown and marketed, tomatoes prefer a warm climate, with ample spacing to encourage abundant yields.

In the North, plants can be sold after all danger of frost has passed until June.

In the southern U.S.: Plant to the garden from February until late April, and again in August for a fall crop. For fall crops, choose varieties with the earliest yield, such as Early Girl (52 days) and Sweet 100 (65 days).

Recommended varieties: Keep in mind that tomato varieties come in 2 classifications—determinate and indeterminate. Determinate indicates that plants grow to the size of small bushes (3 to 4 ft./91.5 to 122 cm.), and stay this height for the remainder of the season. Determinate varieties are often caged with fence wire to keep them upright and the fruit off the ground. Indeterminate types grow

WATERMELON

Ball Culture: In the northern U.S., have plants ready for sale after all danger of frost has passed until late May.

In the southern U.S.: Plant from March to April for spring sales, and again from mid-July to no later than early September for a fall crop.

*Like a number of other vining crops, watermelon does not respond well to transplanting. Sow direct to a Jiffy-pot or similar container where pot and all can be planted. Watermelons are the slowest-growing of all the vining

Crop Time (Weeks)		Number Plants Per Pot	Days To Maturity	Spacing		Gardening	
Cell Packs	Pots 4-in./10-cm.	4-in./10-cm.		In Rows	Between Rows	Height	Width/Spread
—	4–6 (Jiffy)	1–2	90-120	3–4 ft. 91.5–122 cm.	6-8 ft. 1.8-2.4 m.	15–18 in. 38–46 cm.	6–7 ft. 1.8–2.1 m.

In the southern U.S.: Pumpkin plants can be sold from March to late April. Some growers get a second crop from sowing in July for August sales, though this method is probably more practical for Florida and the most southern regions of the U.S.

*Pumpkins do not handle transplanting well and root disturbance reduces the yield. Sow 2 to 4 seeds per container and thin out as needed.

| | 4–6 (Jiffy) | 1–2 | 49-85 | 3–4 ft. 91.5–122 cm. | 5–7 ft. 1.5-2.1 m. | 18–24 in. 46–61 cm. | 3–6 ft. 91.5–183 |

In the southern U.S.: Plants should be sold from late February to late April. In the southern tip of Florida and southern California, zucchini type squash can be planted in August for a winter harvest.

Recommended varieties: For zucchini or summer squash, try Aristocrat (52 days), a green-skinned type, or Sundance (50 days), a yellow, crooked neck variety. Both are F$_1$ hybrids and excellent producers in the garden. For winter squash, try Early Butternut (85 days). Unlike regular Butternut, this variety doesn't require extra space but provides heavy yields.

| | 9–10 | 1–2 | 65–80 | 10–12 in. 25.5–30.5 cm. | 3 ft. 91.5 cm. | 10–12 in. 25.5–30.5 cm. | 12–14 in. 30.5–35.5 cm. |

In the northern U.S., have plants ready for sale once the danger of hard frost has passed until May.

In the southern U.S.: Plant in the fall from late August until late October. Plants may yield in the South after a fall planting or not until spring.

| 6–8 | 8–10 | 1 | 45–96 | 2 ft. 61 cm. | 3–4 ft. 91.5–122 cm. | 3–6 ft. 91.5–183 cm. | 2–3 ft. 61–91.5 cm. |

indefinitely unless trimmed back, and can get as tall as 6 or 7 ft. (1.8 or 2.1 m.). These plants are usually staked on poles.

The most popular indeterminate varieties include: Early Girl (62 days), with 4 to 6-oz. (113 to 170-g.) red fruit; Better Boy (72 days), one of our most popular tomatoes with fruit up to 1 lb. (.45 kg.); Champion (62 days), more disease-resistant than Early Girl and earlier than Better Boy, with 10-oz. (284-g.) red tomatoes; Sweet 100 (65 days), a red variety with loads of 1-in. (2.5-cm.) tomatoes perfect for salads and appetizers; and Heartland (68 days), our most dwarf indeterminate variety, with bright red, 6 to 8-oz. (170 to 227-gm.)

fruit. One of our best determinate varieties is LaRoma (62 days), a productive, good-tasting variety with 3 to 4-oz. (85 to 113-g.) tomatoes great for pasta sauces and canning.

| | 6–7 (Jiffy) | 1–2 | 70–85 | 2–3 ft. 61–91.5 cm. | 5–7 ft. 1.5–2.1 m. | 15–18 in. 38–46 cm. | 7 ft. 2.1 m. |

crops listed in this guide, so count on at least 1 to 2 additional weeks for plants to develop compared to faster-growing squash or pumpkin crops.

Class	Type	Number of Seeds	Germination Temperature		Lighting	Days To Germinate
			Fahrenheit	Celsius		
AGROSTIS A. nebulosa (Cloud Grass)	A, C	266,000/oz. 9,400/g.	70°–72°	21°–22°	C Lt.	3–7

Ball Culture: A quick-developing grass that can be sown direct to the final container and thinned to no less than 2 or 3 seedlings per cell of a tray containing 48 cells per flat. Plants are salable green in 7-8 weeks and will flower shortly thereafter. However, plants are not long-lived in the garden and often die by the middle of July here in our climates. Treat as a dried cut flower and short-lived annual grass.

Class	Type	Number of Seeds	Germination Temperature		Lighting	Days To Germinate
			Fahrenheit	Celsius		
BRIZA B. maxima (Quaking Grass)	A, C	5,600/oz. 200/g.	70°	21°	C. Lt.	4–6
B. minima (Short Quaking Grass)	A, C	93,500/oz. 3,300/g.	70°	21°	C. Lt.	4–6

Ball Culture: The seed heads of both crops are shaped like small hearts, except that B. maxima has slightly larger seed heads than does B. minima.

Sowings made in April will produce lush, green, salable plants 8 weeks later. Using 3 to 4 seedlings per individual cell of a 48 tray will promote faster growth, but 2 seedlings per cell are fine as well. Remember to sow direct to the final container or plug tray and transplant to the garden or a larger container from there.

B. maxima and B. minima are short-lived grasses. If sown in April, cell pack transplants moved into the garden in May will flower in June and the plants will be dead by mid July. Use Briza as a short stemmed cut flower or sow seed direct to pots (5-in.) for unusual pot sales.

Class	Type	Number of Seeds	Germination Temperature		Lighting	Days To Germinate
			Fahrenheit	Celsius		
COIX C. lacryma-jobi (Job's Tears)	A, C	115/oz. 4/g.	70°	21°	C	10–15

Ball Culture: Coix is a large-seeded grass variety which has a gray cast or mottling about the outer seed coat. This plant is grown more as a novelty or curiosity rather than a grass with strong ornamental value.

Sow 1 seed per cell of a 48- or 32- cell pack and the plants will develop and grow quickly. The leaves are the broadest in size of any grass described in this section. Seedlings often resemble corn in their appearance.

In the garden, plants will last until frost but often the leaf tips exhibit a burned edge that has been linked to poor water distribution around the roots. However, the plants will continue to develop and fill in readily.

Class	Type	Number of Seeds	Germination Temperature		Lighting	Days To Germinate
			Fahrenheit	Celsius		
CORTADERIA C. selloana (Pampas Grass)	T. Pe, C	130,000/oz. 5,000/g.	70°	21°	C. Lt.	4–8

Ball Culture: In areas further south, they can reach between 6 and 10 ft. tall. In the northern U.S., use the foliage as an accent or grow the plants in large containers for flower shows in botanical centers if blooms are needed. Sowings should be made directly to the final cell pack or pot. Nine weeks after sowing to a 48-cell tray, the 2-3 plants per cell will not have produced enough foliage to fill out the container, but the roots will be fully wrapped and often have to be trimmed back after 13 weeks.

An early-March sowing will produce salable flats by early to mid May when grown 2-3 seedlings per 48-cell tray. In the northern U.S., plants are not hardy and will die in severe winter areas. A March sowing will seldom produce free- flowering plants in zones 6 and further north unless brought in as divisions from a supplier. Keep in mind that there is another Pampas Grass that is hardier in the northern U.S. but is botanically known as Miscanthus.

In the southern U.S.: Pampas grass is a hardy perennial from zones 8 to 10 but will also be hardy in selected areas further north depending on location and winter protection. Plants will grow to 10 ft. tall once established and are often trimmed back or burned off annually. Plants are often considered weeds but are being used successfully as deciduous shrubs in highway beautification programs across the south. Once established, plants bloom in the summer and fall and the plumes are left in place throughout the cool/cold months to provide winter color and dramatic form. Cortaderia also tolerates salt spray so can be used in beach front plantings.

Growing On Temperature		Crop Time Green Packs (Weeks)	Garden Height	Location	Tender/Hardy
Fahrenheit	Celsius				
55°	13°	7–8	8-10 in. 20–25 cm.	F. Sun	T
58°–60°	14°–15°	7–8	18–20 in. 46–51 cm.	F. Sun	T
58°–60°	14°–15°	7–8	12–15 in. 30–38 cm.	F. Sun	T
58°–60°	14°–15°	8–10	2–3 ft. 60–90 cm.	F. Sun	T
58°–60°	14°–15°	8–10	6 ft.* 1.8 m.	F. Sun	H

Recommended varieties: From seed White Feather and Pink Feather are the two predominant varieties sold by seed houses. The seed is cleaned for easier use and for plug growing.

*Plants will only grow to 3 ft. in the northern U.S. since the growing season is almost over by the time they are large enough to bloom in the fall.

ORNAMENTAL GRASSES

Class	Type	Number of Seeds	Germination Temperature		Lighting	Days To Germinate
			Fahrenheit	Celsius		
FESTUCA F. amethystina (Amethyst Fescue, Large Blue Fescue)	Pe	42,000/oz. 1,480/g.	70°	21°	C. Lt.	6–12
F. ovina glauca (Sheep's Fescue, Blue Fescue)	Pe	42,000/oz. 1,480/g.	70°	21°	C. Lt.	4–6

Ball Culture—F. amethystina: This variety is slower-growing than the common F. ovina glauca but just as attractive. Large blue fescue is sold for its tuft of blue-green foliage rather than for its flowers. F. amethystina has deeper green foliage that lacks the grey cast of F. ovina glauca.

Sowings made in April direct to cell packs or small containers can be thinned to 2-4 seedlings per cell or 3-6 per pot. However, plants do not require thinning and the more seedlings left (within reason), the faster the crop time. Cell packs of this species require a little more time to be salable

green than do those of F. ovina glauca, but the deep green foliage on the small tufts adds an excellent touch to border plantings. Plants bloom the second year from seed. Hardy from zones 4 to 8. One final comment—the first year from seed you may notice that the plants are more green than

HORDEUM H. jubatum (Squirrel's Tail Grass)	A*	17,000/oz. 600/g.	70°–72°	21°–22°	C. Lt.	3–5

Ball Culture: Sow seed anytime 9 weeks prior to selling green in a 48 cell pack. If sown in early March, plants are transplanted green to the garden in mid May.

Plants will reach the seed development stage by the end of June with a full canopy of developing seed heads from the first to the middle of the month of July.

*A number of references would suggest this is a hardy perennial in our northern locations. Though the plants freely flower the first season from seed they often die prior to fall, especially if the seed head is allowed to mature and disperse the seed.

KOELERIA K. glauca	Pe	79,000/oz. 2,800/g.	70°	21°	C. Lt.	3–7

Ball Culture: Similar in appearance to Festuca glauca in foliage color, this plant has a little less blue/gray coloring. Plants will be salable in 8 weeks when seed is sown direct to the final container and thinned to 3 to 5 seedlings per 48-cell tray.

Plants will not flower the same season they were seeded. Koeleria is most often used to add form to the annual or perennial border rather than for its flowers.

LAGURUS L. ovatus (Hare's Tail Grass)	A, C	45,000/oz. 1,580/g.	70°	21°	C. Lt.	4–8

Ball Culture: Lagurus is a fast-growing grass with velvety foliage and plumes. Sowings made in early April and thinned to 1 or 2 seedlings per cell will be salable green within 8 weeks in a 48-cell tray. If growing for cut flower use, we suggest that you sow the seed directly in the ground or from plugs that are transplanted as soon as they are ready.

Plants bloom 14-16 weeks after sowing but should be allowed to bloom in a large pot (6 in.) or within the garden rather than in a small container. Once flowering is completed, plants will start to look tired and will die by the end of August.

ORNAMENTAL GRASSES

Growing On Temperature		Crop Time Green Packs (Weeks)	Garden Height	Location	Tender/Hardy
Fahrenheit	Celsius				
58°–60°	14°–15°	11–15	8–10 in. 20–25 cm.	F. Sun	H
58°–60°	14°–15°	7–8	6–12 in. 15–30 cm.	F. Sun	H

blue-green. The second year plants will color more readily.

Ball Culture—F. ovina glauca: An excellent perennial grass with grey-green foliage on plants which grow no more than 6 to 7 in. tall the first season after sowing. Festuca will bloom the second year and, when they die, turn golden brown above the uniform tufts of foliage.

Sowings made direct to the final container (cell packs or pots) will produce salable plants that can be transplanted to the garden or into larger containers for holding over the summer. Seed sown in early April and thinned to 2-4 seedlings per cell pack will produce salable green flats 7-8 weeks later.

55°	13°	8–9	20–24 in. 51–61 cm.	F. Sun	T

55°–58°	13°–14°	7–8	10–12 in. 25–30 cm.	F. Sun	H

58°–60°	14°–15°	6–7	18–22 in. 46–56 cm.	F. Sun	T

ORNAMENTAL GRASSES

Class	Type	Number of Seeds	Germination Temperature		Lighting	Days To Germinate
			Fahrenheit	Celsius		
PENNISETUM P. setaceum* (Crimson Fountain Grass, Fountain Grass)	T. Pe, C	49,000/oz. 1,730/g.	70°	21°	C. Lt.	3–6
P. villosum (Feathertop)	T. Pe, C	14,000/oz. 495/g.	70°	21°	C Lt.	3–6

Ball Culture–P. setaceum: Sowings made in spring will flower readily later during the summer and produce blooms up until frost. The light crimson-tipped plumes measure up to 6 in. long on gracefully arching foliage to between 3 and 4 ft. tall. An April sowing produces blooming plants in late June or July with full bloom expected 2-3 weeks later.

Sowings made direct to cell packs (48 cells per flat) and thinned to 2-3 seedlings per cell will be salable green within 8 weeks. Do not attempt to flower within the pack because the roots become large within a short period of time and require cutting to remove the plants from the cell pack. As an example, we sowed seed direct to a 32-cell tray in February, thinning to 2-3 seedlings per cell. We planned on planting out in late May or early June. On June 4th we had to cut all the roots off under the cell packs since they measured up to 12 in. long. Plants had about 10 in. of growth on top and quickly took root in their garden bed location.

Pennisetum setaceum is hardy within zones 8 to 10 and selected areas of 7. They are treated as annuals in zones 6 and above.

*Botanical names are a source of confusion for this plant. It is sometimes sold as P. rueppelii, P. rueppelianum or P. setaceum var. rueppelii. Pennisetum setaceum is one of the best overall grasses from seed for the landscape.

Class	Type	Number of Seeds	Germination Temperature		Lighting	Days To Germinate
			Fahrenheit	Celsius		
PHALARIS P. canariensis (Canary Grass)	A, C	3,600/oz. 127/g.	70°	21°	C. Lt.	4–8

Ball Culture: This easy-to-grow grass is short-lived in the garden. Sow seed direct to the final cell pack, small container, or garden and do not attempt to transplant individual seedlings. When seeding to cell packs, sow so as to have 2 or 3 finished seedlings per individual cell of a 48-cell tray.

Plants are quick to develop and will be salable green 6 to 7 weeks later. Do not be tempted to hold onto the flats once they are 7 to 8 weeks old. If left for up to 9 weeks plants often flower under stress, spoiling later performance. If you plan on holding onto the plants for some time, it would be wise to sow direct into a larger container as opposed to a small cell pack. If sown in April and transplanted to the garden in mid May, plants will flower in June and July and will die by mid August.

If growing for use as a cut flower, keep in mind that the green seed heads are excellent as a fresh cut filler, while the brown seed heads can be dyed and used in fall or colonial accent arrangements.

ORNAMENTAL GRASSES

Growing On Temperature		Crop Time Green Packs (Weeks)	Garden Height	Location	Tender/Hardy
Fahrenheit	Celsius				
58°–60°	14°–15°	7–8	3–4 ft. 91–122 cm.	F. Sun	T
58°–60°	14°–15°	7–8	2.5–3 ft. 61–91 cm.	F. Sun	T

Ball Culture–P. villosum: Unlike P. setaceum, P. villosum bears cream-white plumes that are plump and only 4 in. long. Seeded forms of this grass can be weak and collapse after being watered overhead or rained on. Cultivars of this selection that are vegetatively propagated have fared much better with a stronger landscape performance overall than the seeded types.

Sowings made in the spring will be salable green 7-8 weeks later in a 48-cell flat with 2 or 3 seedlings per cell. Although the plants are perennial, they are classified as a tender perennial because they will not survive a northern winter.

58°–60°	14°–15°	6–7	2–3 ft. 61–91 cm.	F. Sun	T

ORNAMENTAL GRASSES

Crop Index